Power Supply Projects

The Maplin series

This book is part of an exciting series developed by Butterworth-Heinemann and Maplin Electronics Plc. Books in the series are practical guides which offer electronic constructors and students clear introductions to key topics. Each book is written and compiled by a leading electronics author.

Other books published in the Maplin series include:

Computer Interfacing	Graham Dixey	0 7506 2123 0
Logic Design	Mike Wharton	0 7506 2122 2
Music Projects	R A Penfold	0 7506 2119 2
Starting Electronics	Keith Brindley	0 7506 2053 6
Audio IC Projects	Maplin	0 7506 2121 4
Auto Electronics Projects	Maplin	0 7506 2296 2
Video and TV Projects	Maplin	0 7506 2297 0
Test Gear & Measurement	Danny Stewart	0 7506 2601 1
Integrated Circuit Projects	Maplin	0 7506 2578 3
Home Security Projects	Maplin	0 7506 2603 8
The Maplin Approach to Professional Audio	T.A. Wilkinson	0 7506 2120 6

Power Supply Projects

BH NEWNES

Newnes

An imprint of Butterworth-Heinemann Ltd

Linacre House, Jordan Hill, Oxford OX2 8DP

A member of the Reed Elsevier group

OXFORD   LONDON   BOSTON
MUNICH   NEW DELHI   SINGAPORE   SYDNEY
TOKYO   TORONTO   WELLINGTON

British Library Cataloguing in Publication Data
A catalogue record for this book is available from the
British Library
ISBN 0 7506 2602 X

Library of Congress Cataloguing in Publication Data
A catalogue record for this book is available from the
Library of Congress

Edited by Co-publications, Loughborough

Typeset and produced by Sylvester North, Sunderland

all part of The Sylvester Press

Printed in Great Britain by Clays Ltd, St Ives plc

Contents

Preface

This book is a collection of projects previously published in *Electronics — The Maplin Magazine*.

While all circuits given with all integrated circuits here are intended for experimental use only — they are not full projects by any means — a printed circuit board track and layout are detailed. To help readers, the printed circuit boards (and kits of parts for some of the projects, too) are available from Maplin, but — as these circuits *are* for experimental use only — all constructional details and any consequent fault finding is left upto readers.

This is just one of the Maplin series of books published by Newnes books covering all aspects of computing and electronics. Others in the series are available from all good bookshops.

Maplin Electronics Plc supplies a wide range of electronics components and other products to private individuals and trade customers. Telephone: (01702) 552911 or write to Maplin Electronics, PO Box 3, Rayleigh, Essex SS6 8LR, for further details of product catalogue and locations of regional stores.

1 Laboratory power

Low cost PSU

This low cost power supply is a relatively simple design that provides reliable performance and is ideal as a power supply for the home constructor. The supply makes available a variety of voltage combinations which include variable split supply, variable single supply, a fixed 5 V and a fixed 12 V supply. A three position, switchable current limit is also provided and the unit is capable of supplying current levels up to 1 A.

Circuit description

Figure 1.3 shows the circuit diagram of the power supply. The supply is based around the L200C regulator, which is capable of handling input voltages up to a maximum of 40 V and output voltages up to 30 V with programmable current limit. The circuit effectively uses two individual single power supplies i.e. a transformer with two separate secondary windings and two regulators.

A 2 A transformer is used in the design to allow plenty of headroom when the power supply is being used at a current level of 1 A. It is important that the transformer secondary voltage is not allowed to drop below the minimum input voltage for full voltage output from the regulator, as regulation would be lost.

Mains voltage is applied to transformer T1 via primary fuse FS1 and mains on/off switch S1. A mains voltage of 240 V r.m.s. on the primary of the transformer, corresponds to a secondary voltage of approximately 20 V r.m.s. The low voltage a.c. is fed to two separate bridge rectifiers (BR1 and BR2) via fuses FS3 and FS2. The output from the rectifiers is smoothed by electrolytic capacitors C1 and C2. Two completely separate unregulated d.c. supplies are produced which are then individually fed to the input of regulators IC1 and IC2. The output voltage of each regulator is determined separately by a network of resistors, switched by S3. A similar set of resistors (switched by S2) are used for current limiting purposes. S3 is also used to select single or split supply operation. Transistors TR1 and TR2 are used to drive two LEDs which indicate the status of the power supply outputs (single or split). Diodes, D1 and D2 are used to prevent voltage spikes or residual voltages from external equipment connected to the output of the supply from damaging the regulators. Fast recovery diodes are used in this application because of their fast switching characteristics. Resistors R7 and R15 ensure that the diodes are maintained in a conducting state even at very low output current levels. Capacitors C3 and C4 decouple the regulators, attenuate noise and prevent instability.

Construction

Insert and solder the components onto the PCB referring to Figure 1.2 and the Parts List. It is a good idea to start with the resistors, as these are relatively low profile components, and may be awkward to fit at a later stage. Next, using the resistor lead off-cuts, fit the eight links on the PCB, these are marked *link* on the legend. The fuse clips (used to hold FS1 and FS2) should then be fitted; these must be kept flush with the PCB when soldering as illustrated in Figure 1.1. Next insert and solder

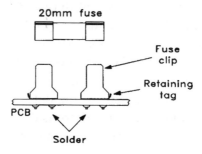

20mm fuse

Fuse clip

Retaining tag

PCB

Solder

Figure 1.1 Mounting the fuse clips

the capacitors. It is important that the electrolytic capacitors are fitted with the correct polarity; the negative lead of the capacitor, marked by a negative (–) symbol on the component body, is inserted into the hole furthest from the positive (+) symbol on the legend. Transistors TR1 and TR2 are fitted such that the case of the component corresponds with the outline on the PCB legend. Regulators RG1 and RG2 are fitted in a similar manner with the heatsink tags perpendicular to the PCB; the tags are bolted down to the bottom of the case, using insulating bushes and washers, when the PCB is finally installed. Bridge rectifiers, BR1 and BR2 are inserted and soldered such that the symbols on the corners of the device correspond with those on the PCB legend. The bridge rectifiers each use a small heatsink and these are held in place by a nut and bolt through the PCB. Position the heatsink such that it is clear of any surrounding components. Potentiometers RV1 and RV2 are mounted on the component side of the PCB as shown in Figures 1.4(a) and 1.4(b). Switches S2 and S3 are mounted in a similar manner to RV1 and RV2 and are connected to the PCB using insulated hook-up wire as shown in Figure 1.5. Fit the two LEDs on the track-side of the PCB, ensuring correct orientation, this is indicated by the dotted legend on the component-side of the PCB. Finally fit the two track-side links using insulated hook-up wire. The location of each, is indicated on the component-side of the PCB by a dotted line; a circle at each end of the dotted lines indicates the position where the wire ends should be soldered to the track. Please note, to prevent instability these links must be fitted exactly where indicated by the legend and not at any other point.

For further information on soldering and construction techniques, reference should be made to the constructors' guide included in the kit.

Enclosure and wiring

Before the supply can be powered up, the transformer, PCB and associated components *must* be housed in a suitable metal case. The recommended case is Steel Case 1608 (stock code XJ28F) and the drilling details, for those wishing to use this case, are shown in Figure 1.6. PCB mounting information is shown in Figure 1.7.

For ease of assembly, it is recommended that the wiring to the PCB is made before fitting into the case, and that the PCB is fitted before the transformer. With the recommended case, it may be found advantageous to remove the rear panel when installing the PCB and transformer.

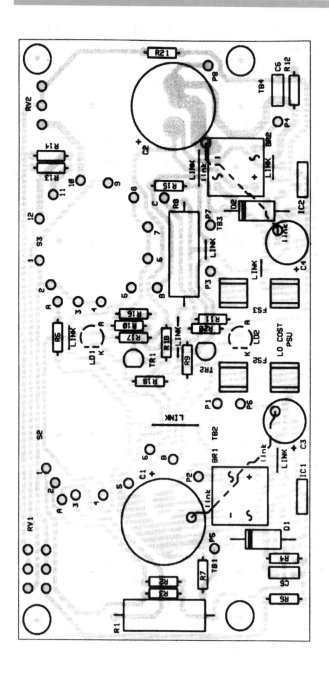

Figure 1.2 PCB legend and track

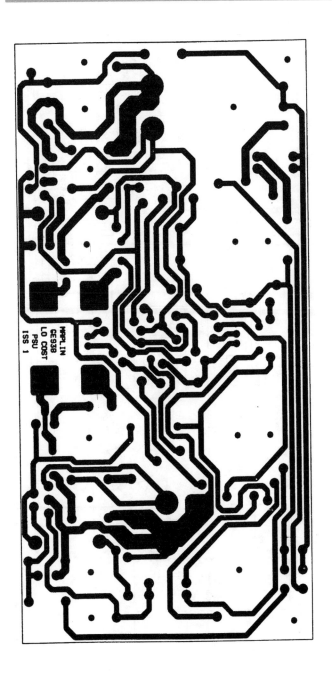

Figure 1.2 Continued

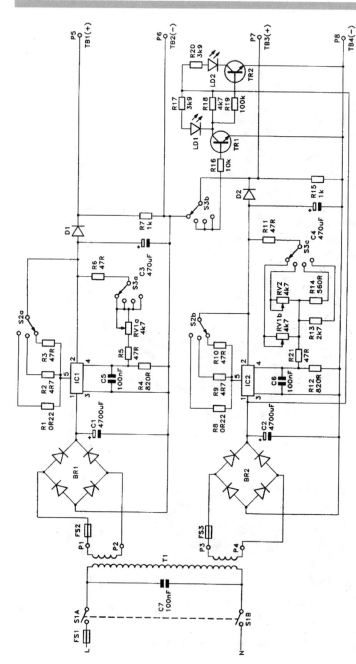

Figure 1.3 Circuit diagram

Laboratory power

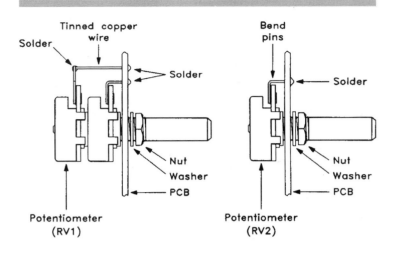

Figure 1.4   (a) Mounting RV1, (b) mounting RV2

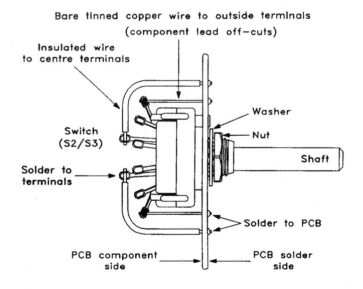

Figure 1.5   Mounting S2 and S3

9

Power supply projects

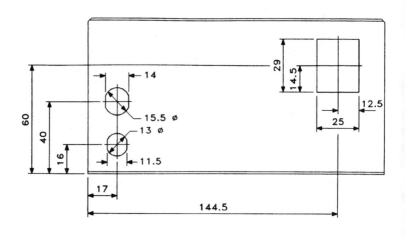

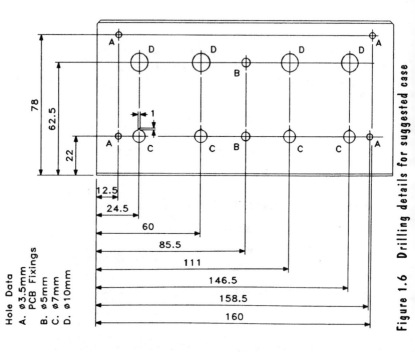

Hole Data

A. ∅3.5mm PCB Fixings
B. ∅5mm
C. ∅7mm
D. ∅10mm

Figure 1.6 Drilling details for suggested case

10

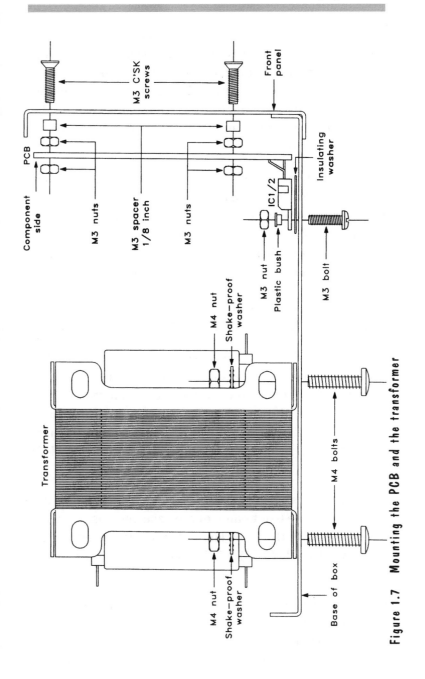

Figure 1.7 Mounting the PCB and the transformer

Power supply projects

It is important that the case used has no large holes as live mains is present inside, on the primary side of the transformer, and it is important that the risk of electric shock due to touching any of these parts is eliminated. Figure 1.8 shows connections to the mains fuse holder (FS1), on/off switch, suppression capacitor and transformer (T1) primary.

Transformer secondary (low voltage) wiring is shown in Figure 1.9. All mains leads should be shrouded using heatshrink sleeving. The wiring should be double checked to make sure that there are no errors. Connection of the earth lead is an essential safety precaution; make sure that the earth lead (colour coded green/yellow) is securely connected to the tag provided, and that the tag is bolted securely to the chassis, so as to make a good electrical connection.

The front panel layout is basically determined by the position of the front panel components, but the actual front panel legend is down to the user. A suggested front panel layout is shown in Figure 1.10; the illustration also shows the different switch positions and approximate voltage settings for the variable voltage controls (RV1 and RV2).

Terminal posts TB1 (+), TB2 (–), TB3 (+) and TB4 (–) are mounted on the front panel of the case and pass through 4 large holes in the PCB. The tags are connected to PCB pins P5 (TB1), P6 (TB2), P7 (TB3) and P8 (TB4) as shown in Figure 1.11.

It is important that RG1 and RG2 are suitably heatsinked. The metal case suggested will provide sufficient heatsinking under normal conditions. However, if a

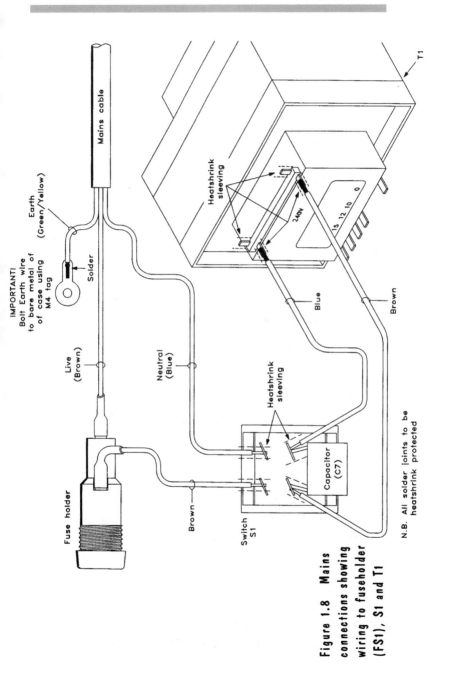

Figure 1.8 Mains connections showing wiring to fuseholder (FS1), S1 and T1

Power supply projects

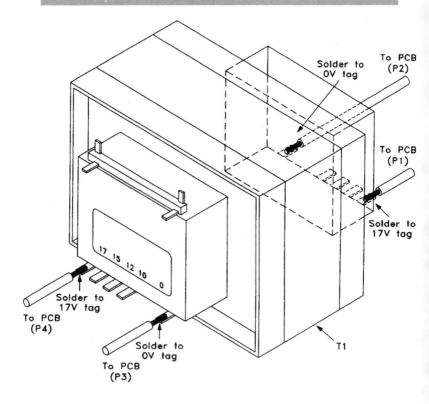

Figure 1.9   Transformer secondary connections

different case is used, it may be necessary to provide additional heatsinks. Failure to provide the rated current level due to the regulators *shutting down* suggests additional heatsinking is necessary.

Because the bottom of the case is used as a heatsink, it is important that the case is slightly raised to allow air to flow freely under the case-bottom. Four rubber feet (one at each corner) can be used for this purpose.

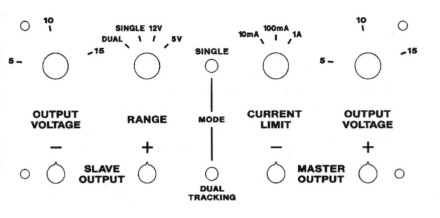

Figure 1.10   Front panel layout showing switch positons and approximate voltage settings

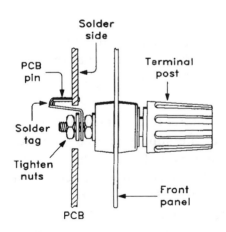

Figure 1.11   Terminal post connection

15

Power supply projects

Testing

Before the circuit is powered up it is essential that the transformer, PCB and associated circuitry is installed in an earthed metal case, and the cover or lid is firmly in place; this eliminates the risk of electric shock by accidentally touching live parts. *Important: any work on the circuit should be carried out with the mains supply disconnected*, and the supply should *never* be powered up with the cover of the case removed. Remember, *mains voltage can kill.*

It is recommended that you double-check your work before powering up the supply, to make sure that there are no dry joints or solder short circuits, and to ensure that all off-board wiring is correct.

Insert the fuses into the fuseholders; the $1^1/_4$ inch fuse (FS1) is inserted into the chassis fuse holder, which should be mounted on the rear panel, and the two 20 mm fuses (FS2 and FS3) are fitted into the appropriate fuse clips on the PCB.

The circuit requires no alignment and once construction is complete, the supply should be ready for use. A multimeter is required to test the supply properly. Initially, rotate RV1 and RV2 fully anti-clockwise. Set current limit switch S2 to the 1 A position and voltage range switch S3 to the tracking position (position 1). With S3 in this position, the supplies are coupled together, and an output voltage of approximately 3 V should be present between each set of outputs (TB1 and TB2, and TB3 and

TB4); the voltage between TB1 and TB4 should be around 6 V. In the tracking mode, LD2 should light. As RV1 is rotated clockwise the voltage should increase to at least 15 V, between each set of outputs, and approximately double this between TB1 and TB4. The voltages stated may vary somewhat due to component tolerances. Set S3 to the variable single position (position 2). Similar results should be obtained, but this time RV1 controls the voltage between TB1 and TB2, and RV2 controls the voltage between TB3 and TB4. The voltage between TB1 and TB4 should be 0 V on this range, as the supplies are separate and are not directly coupled together. In all modes except the tracking mode, LD1 should light indicating single supply operation. With S3 in the 12 V position (position 3), a fixed 12 V supply is available between TB3 and TB4, and a variable supply between TB1 and TB2. With S3 in the 5 V position (position 4), a fixed 5 V supply is available between TB3 and TB4, with a variable supply between TB1 and TB2. Table 1.1 summarises the different output voltage combinations available and Table 1.2 shows the three different current limit settings. The voltage range and current limit threshold figures shown are only approximate, and these may vary considerably. Switch S2 sets the current limit for both sets of outputs. The actual current limit thresholds may be measured by connecting a multimeter, set to measure current, between TB1 and TB2 or TB3 and TB4, such that the supply is temporarily short-circuited. This procedure should be repeated on each of the three current ranges. To prevent stress on the components, it is recommended that the power supply is not left in a short circuit condition for more than 1 minute, although under normal conditions the supply should be capable of operating into a short circuit for considerably longer.

Power supply projects

S3 position	TB1 and TB2	TB3 and TB4
1	Variable tracking Master (RV1), 3 V to 15 V	Variable tracking Slave (RV1), 3 V to 15 V
2	Variable single (RV1) 3 V to 15 V	Variable single (RV2) 3 V to 15 V
3	Variable single (RV1) 3 5 to 15 V	12 V fixed (RV2)
4	Variable single (RV1) 3 V to 15 V	5 V fixed (RV2)

Table 1.1 Output voltage ranges

S2 Position	Approximate current limit threshold
1	10 mA
2	100 mA
3	1 A

Figure 1.2 Approximate current limit thresholds

Using the power supply

The power supply is suitable for general-purpose use and should provide a relatively smooth regulated output, if the guidelines set out in this chapter are adhered to. Table 1.3 gives specifications of the prototype power supply. The outputs are protected against unwanted positive transients and the supply features full overload protection.

The supply outputs should exhibit very few unwanted switching transients; however, to prevent any possible damage it is recommended that the power supply is initially switched on *before* any loads are connected.

To prevent overheating, the supply should not be located where the free-flow of air is inhibited as this will impede heat dissipation.

In the dual tracking mode (S3 in position 1) TB2 is connected to TB3 internally and an external link is not required for this purpose. In this mode TB1 is the +V output, TB2 or TB3 are 0 V connections and TB4 is the –V output. In all other modes two completely separate supplies are provided; one set of outputs is available from TB1 (+V) and TB2 (–V) and a separate set is available at TB3 (+V) and TB4 (–V).

Input voltage to transformer (T1)	240 V a.c. mains
Input voltage to power supply PCB (off load)	23 V r.m.s.
Power supply output voltage	see Table 1.1
Maximum output current	see Table 1.2
Output ripple voltage (500 mA output current)	5 mV

Table 1.3   Specification of prototype

Low cost power supply parts list

Resistors — All 0.6 W 1% metal film [unless specified]

R1,8	0.22R W/W min	2	(W0.22)
R2,9	4R7 min res	2	(M4R7)
R3,5,6, 10,11,21	47R min res	6	(M47R)
R4,12	820R min res	2	(M820R)
R7,15,16	10 k min res	3	(M10K)
R13	3 k min res	1	(M3K)
R14	750R min res	1	(M750R)
R17,20	3k9 min res	2	(M3K9)
R18	4k7 min res	1	(M4K7)
R19	100 k min res	1	(M100K)
RV1	4k7 dual pot lin	1	(FW84F)
RV2	4k7 pot lin	1	(FW01B)

Capacitors

C1,2	4700 uF 35 V PC elect	2	(JL30H)
C3,4	470 uF 35 V PC elect	2	(FF16S)
C5,6	0.1 uF 50 V disc	2	(BX03D)
C7	0.1 uF IS cap	1	(JR34M)

Semiconductors

IC1,2	L200CV	2	(YY74R)
TR1,2	BC548	2	(QB73Q)
D1,2	BYW98-150	2	(UK65V)
LD1	2 mA 5 mm LED red	1	(UK48C)
LD2	22mA 5 mm LED green	1	(UK49D)
BR1,2	S04	2	(QL10L)

Laboratory power

Miscellaneous

T1	68VA 172 V multi	1	(WB22Y)
S1	DPST rocker	1	(YR69A)
S2	SW3B rotary	1	(FF76H)
S3	SW4 rotary	1	(FH44X)
FS1	160 mA 1 1/4 A/S fuse	1	(UJ99H)
FS2,3	2 A A/S fuse	1	(WR20W)
	1 1/4 clickcatch F/H	1	(FA39N)
	5R2 SR grommet	1	(LR48C)
	pin 2141	1	(FL21X)
	large term post blk	2	(HF02C)
	large term post red	2	(HF07H)
	20 mm fuse clip type 1	4	(WH49D)
	CP 32 heat shrink	1	(BF88V)
	slotted heatsink	2	(FL58N)
	TO220 insulator	2	(QY45Y)
	TO66 bush	1	(JR78K)
	16/0.2 wire 10 M red	1	(FA33L)
	M4 12 mm isobolt	1	(BF49D)
	M4 isonut	1	(BF57M)
	M4 isotag	1	(LR63T)
	M3 16 mm isobolt	1	(JD16S)
	M3 isonut	1	(BF58N)
	lo cost PSU PCB	1	(GE93B)
	lo cost PSU leaflet	1	(XT05F)
	constructor guide	1	(XH79L)

Optional (not in kit)

1608 steel case	1	(XJ28F)
min mains black	1	(XR01B)
M4 12 mm steel screw	1	(JY15R)
M4 steel nut	1	(JD60Q)

Power supply projects

K7A knob	2	(YX01B)
K7B knob	2	(YX02C)
M4 isoshake	1	(BF43W)
M3 isonut	1	(BF58N)
M3 isoshake	1	(BF44X)
$^1/_8$ in M3 spacer	1	(FG32K)
M3 16 mm pozi screw	1	(JC70M)

The above parts (excluding optional) are available as a kit, order as LP74R

A versatile power supply

We can't do very much experimenting in electronics without suitable sources of d.c. voltage. While most TTL digital circuits require only a +5 V supply at modest current and many operational amplifiers will be happy with a balanced ±15 V supply, a variety of other circuits need anything from about 3 V to 20 V or more. This power supply design caters for all of the above cases and details are given for its construction. IC regulators are used throughout, thus minimising the component count and cost. The complete circuit diagram for this power supply is shown in Figure 1.12.

The circuit uses a miniature mains transformer with a dual secondary arrangement. These latter windings are connected to a bridge rectifier BR1 so as to produce a pair of unregulated d.c. outputs of equal value but opposite polarity. These voltages appear across the large value electrolytic capacitors C1 and C2 respectively. The nominal values of these voltages appear on the circuit for reference. The positive d.c. output feeds all three regulators, while the negative supply is used only for the 4195 regulator in order to develop the −15 V output.

The case specified will accommodate the circuit components adequately. Some flexibility has been left in the actual physical arrangements of the unit. It is stressed, however, that this project should only be undertaken when the right level of competence in soldering, etc., has been attained and/or supervisory help from a more qualified person is available. Particular care should be taken with the mains wiring of the unit; the mains

Power supply projects

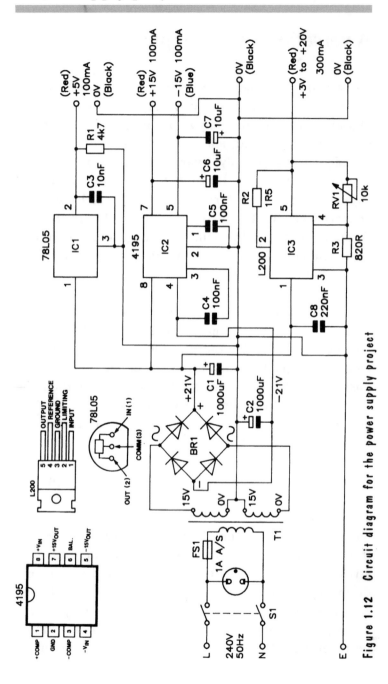

Figure 1.12 Circuit diagram for the power supply project

cable should be securely anchored to relieve stress on ~~the transformer connections~~.   correct

but it does not connect with the T. (Sw + fuse first!)

Testing the completed unit

When wiring is complete, a visual check should be carried out to detect any errors, bad joints, etc. Make sure that the electrolytic capacitors are the right way round and that the rectifier and transformer are correctly connected. Particular care always has to be taken in the case of mains-operated equipment. When satisfied that all ought to be well, the time has come to switch on. If this act produces nothing other than a comforting silence, a voltmeter should be used to check the outputs of the +5 V and ±15 V sections. Then the variable output should be checked, giving an approximate range from 3–20 V. The variable control should, logically, produce an increasing voltage as it is rotated clockwise. If all is well, the unit can be boxed up, the terminals marked with transfer lettering and marking to indicate the outputs and you now have your own *hand-built* power supply ready for use. Perhaps just the first of many such items. Apart from saving substantially on the cost of factory made equipment, there is the satisfaction aspect, as well as the increasing self-confidence.

Appendix

Energy stored in a capacitor

The capacitor is able to store energy in the form of an electrostatic field between its plates. While it is storing

Power supply projects

energy, a voltage exists between its terminals. The amount of stored energy is calculated from the energy W (Joules) stored in a capacitor of C (Farads) with a terminal voltage V (Volts) is given by.

$$W = (1/2) \times C \times V^2$$

Example: a 10 µF capacitor with a terminal voltage of 20 V is storing

$$0.5 \times 10 \times 10^{-6} \times 20^2$$
$$:= 0.5 \times 10^{-5} \times 400$$
$$= 2 \, mJ \, (milli - Joules)$$

Energy stored in an inductor

The inductor (coil) is able to store energy in the form of an electromagnetic field within its turns. While it is storing energy, a current is flowing through it. The amount of stored energy is calculated from the energy W (Joules) stored in an inductor of L (Henries) with a current of I (Amperes) flowing is given by:

$$W = 0.5 \times L \times I^2$$

Example: a 20 H coil carrying a current of 5 A is storing:

$$0.5 \times 20 \times 5^2$$
$$= 0.5 \times 20 \times 25$$
$$= 250 \, J \, (Joules)$$

Power supply parts list

Resistors

R1	4k7	1 (M4K7)
R2	1.5 Ω	1 (M1R5)
R3	820 Ω	1 (M820R)
RV1	10 k lin pot	1 (FW02C)

Capacitors

C1,2	1000 µF 35 V axial electrolytic	2 (FB83E)
C3	10 nF disc ceramic	1 (BX00A)
C4,5	100 nF disc ceramic	2 (BX03D)
C6,7	10 µF 25 V axial electrolytic	2 (FB22Y)
C8	220 nF disc ceramic	1 (JL01B)

Semiconductors

IC1	78L05	1 (QL26D)
IC2	4195	1 (XX02C)
IC3	L200	1 (YY74R)

Miscellaneous

T10 transformer–15 V; 0– 15 V 10 VA	1 (LY03D)
bridge rectifier BR1 type W01	1 (QL38R)
case	1 (LQ09K)
S1 switch	1 (YR70M)
20 mm fuseholder	1 (RX96E)
1 A A/S fuse	1 (WR19V)

Power supply projects

knob	1 to suit
4 mm socket red	3 (HF73Q)
4 mm socket black	3 (HF69A)
4 mm socket blue	1 (HF70M)

Laboratory power supply

This extremely flexible laboratory power supply unit (PSU) is capable of sourcing well-regulated d.c. voltages of up to 30 V at currents of up to 8 A continuous, 10 A peak. As a result, there is a wide range of potential applications for the hobbyist, service department and educational institutions. For example, it is ideal for the testing of prototypes; in addition to the sheer power output available, there is a current-limiting function — ideal for trying out your more delicate circuits. This function, and the robust nature of this unit, makes it an ideal choice for servicing d.c. equipment (e.g., portable audio and video equipment) and for college workbenches. In addition, this piece of equipment is ideal for the running of CB and amateur radio equipment, and even the charging of batteries; lead-acid packs in *constant voltage* mode, and Ni-Cad cells in *constant current* mode.

Circuit descriptions

Control PCB

(Refer to circuit diagram of Figure 1.13). IC2 and 3 are μA723 voltage/current regulators. IC2 is used to set the output voltage; R6 providing feedback to compensate for the voltage drop across T2, R16, T3 and R31. A fraction of the output voltage determined by R5 and R23 is supplied to IC2 as feedback. IC3 is responsible for current-limiting; its operation is very interesting. A user-set reference voltage (the *current limit*) is derived from

Power supply projects

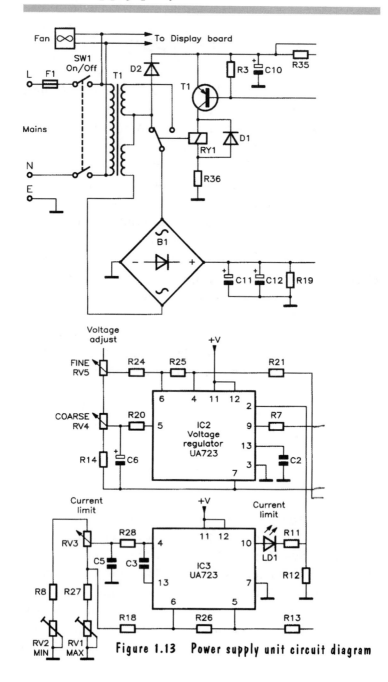

Figure 1.13   Power supply unit circuit diagram

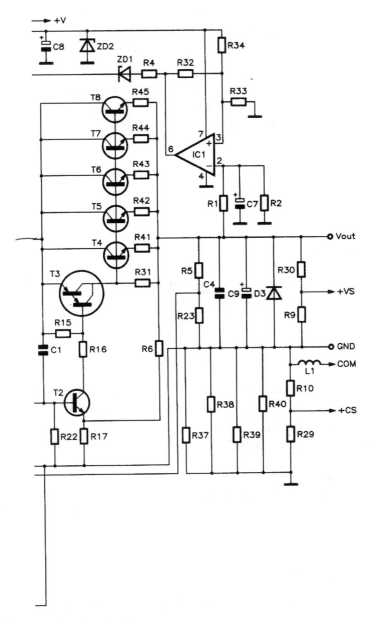

Figure 1.13   Continued

31

Power supply projects

IC3's on-chip reference via RV1/2/3, R18, R27 and R28. This is compared to the voltage developed across R37 to R40 (which are situated between output ground and the reference ground); if it exceeds the user-set reference, then the error voltage produced by IC3 will be sufficient to forward-bias the LED, in doing so it will turn on the current-limiting transistor integral to voltage regulator IC2.

Transistor T2 is the control transistor for T3, a Darlington device, which provides plenty of drive current for the output pass transistors T4 to T8, which are fed with the smoothed d.c. output from the bridge rectifier and reservoir capacitors. Note that the main power transformer is a very beefy (300 VA, 15–0–15 V) toroidal type. R41 to R45 are required to allow for the variation in current gain among the transistors; these items are not matched. Diode D3 is present to protect the power supply from any reverse-polarity voltages that may accidentally be applied to its output terminals.

Integrated circuit IC1 is a 741 op-amp configured as a comparator; it switches in the second winding of the toroidal transformer, via T1 and RY1, when the output voltage rises above 12 V; this voltage is determined by R1/R2.

+VS, +CS and COM are (respectively) the positive voltage sense, positive current sense, and common outputs for the display PCB.

Display PCB

(Refer to circuit diagram of Figure 1.14). Integrated circuits IC1 and IC2 are complete digital voltmeters,

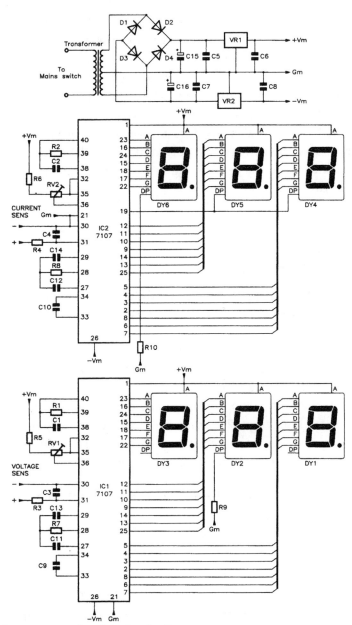

Figure 1.14   Display circuit diagram

Power supply projects

containing all the necessary analogue-to-digital converters and 7-segment display driver/buffers, requiring only a few additional components to function. The display PCB is powered from a split rail (–5 V) supply by VR1/VR2 and associated components. Note that the power supply for the display PCB is derived from a separate transformer. IC2 is used as an *ammeter*; in reality, it measures the voltage (via potential divider R10/R29) that the measured current produces across shunt resistors R37 to R40.

PCB construction

Construction of the four PCBs is fairly straightforward, and instructions are given in the manual supplied with the kit. The most important things to watch out for are misplaced components; it is important that you check the orientation and positioning of a component before soldering it in; desoldering is time-consuming and could lead to damage of the component or board. In particular, watch out for polarised components (such as semiconductors and electrolytic capacitors).

When building the PCBs, there are a few points to watch out for.

Main PCB

Before fitting RV3, 4 and 5 to the PCB, shorten their shafts to 30 mm using a hacksaw to cut off the excess length. Note that these potentiometers should be securely installed on the PCB before making connections to the PCB with tinned copper wire. The same applies to T3; this item (a BD646 transistor), along with its heatsink, should be bolted to the board before soldering its leads in place. To aid the transfer of heat, heatsinking compound (not supplied in the kit) should be smeared onto the tab of the transistor before mounting it. Note that all three ICs on this board are socketed. R37 to R40, the shunt resistors, are made from nichrome resistance wire supplied in the kit. Each is formed from a 10 cm length of the wire, as shown in Figure 1.15. The next point to bear in mind is that the current limit indicator needs to be installed on the *track* side of the board, so that the LED's tip protrudes 30 mm above the board's surface. Note that the two-pole screw connectors J1 to J5 (used to connect the transformer's secondary windings to the PCB) clip together (start with J1) before fitting and soldering in place. After soldering the relay in place, coat the already plated tracks with more solder, so that they can cope with the potentially high currents expected.

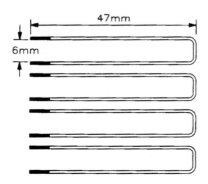

47mm

6mm

Figure 1.15   Forming R37 to R40

Power supply projects

Display PCB

Note that IC1 and IC2 are socketed. In addition, the 7-segment displays are also fitted in sockets — ensure correct polarity when finally fitting these in place. VR1 and VR2 should be bolted into place before soldering them to the PCB. Two electrolytic capacitors (C15 and C16) must be fitted to the solder side of the board, paying special attention to polarity.

Regulator transistor PCBs

Two boards, which fit in the side-mounted heatsinks of the power supply, hold the connections to the TIP3055 transistors (T4 to T9) and their emitter resistors. The two supplied PCBs are identical; however, one should be made to accommodate the connections to two transistors only. On this board, the space marked *T6*, and the adjacent holes for *Lucar* terminals ($^1/_4$ in blade terminals), should be left unpopulated; the space on this PCB directly above on the heatsink is reserved for the bridge rectifier. Do *not* solder the transistors directly to the boards — fit PCB pins in the appropriate positions (marked b, c and e). The corresponding emitter resistors (R41 to R45) should be fitted vertically to the board. Finally, mount the Lucar terminals, and solder into position.

After completing assembly of the boards, check your work thoroughly. Spotting any mistakes before powering up, could save any (possibly expensive!) problems later on.

Mechanical assembly

Detailed assembly instructions for the case are given in the comprehensive manual supplied with the kit. Assembly is straightforward, but a few remarks made here will save trouble later.

Care should be exercised when installing the regulator pass transistors T4 to T8. These, as already mentioned, are connected via PCB pins to two boards located in the unit's side-mounted heatsinks. To fit, slide the boards into the heatsinks (the correct way up and around — the outline of the transistors is screen-printed on the board to aid you). Looking from the front of the unit, the two-transistor board is fitted to the left heatsink, while the other is fitted to the right one. The transistors are screwed to the heatsink using M3 hardware (an insulating washer and heatsinking compound must be used for each device!), and are then soldered to the corresponding PCB pins. Note that heatsinking compound is also used when attaching the bridge rectifier to the left heatsink. These two heatsinks form the sides of the case.

When installing the fan, ensure that the arrow (embossed on its side) points to the back of the unit's case. This indicates airflow direction; in other words, the air is being sucked through the power supply. Air comes in from slots stamped in the top and bottom lids of the case — air flow requirements should be borne in mind when using the finished unit. Do not obstruct the ventilation grilles. Note that the fan should be orientated so that its leads appear at the top. It is fixed to the rear

Power supply projects

panel with 35 mm bolts; if these are over-tightened, the fan's housing could crack. *Finger tightness* plus half a turn should be sufficient.

The mains chassis plug and fuse-holder are the next items to be installed on the rear panel. These are mounted using zinc-plated M3 bolts. Sleeving (not supplied in the kit) should be used to cover all mains connections; insulating boots are available for the chassis plug and fuseholder — see Optional Parts List. Note the earthing arrangements; as shown in Figure 1.16, one of the PCB mounting posts, the power supply case and

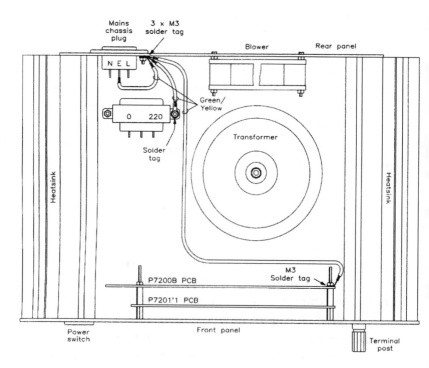

Figure 1.16 Earthing arrangements

the display transformer are all connected to earth. M3 solder tags are used for the bolted components. Note that the paint *must* be scraped away around the chassis plug's screw hole, on the inside of the back panel, to ensure a good earth connection.

The same applies to one of the display transformer mounting screws on the underside of the power-supply; this item should be fitted as shown in Figure 1.17.

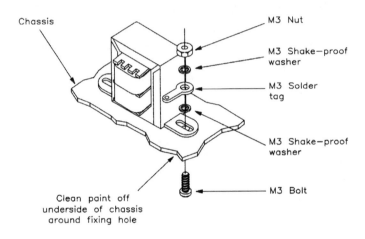

Figure 1.17   Display transformer

The two main PCBs are attached to the main panel by four 45 mm long countersunk M3 bolts. The paint around the top-right M3 hole should be removed; this will eventually form an earth connection. The bolts are held in place on the other side of the panel by threaded hexagonal spacers. These should be fully tightened, as the screw-heads themselves will eventually be covered by the front panel label. After this has been done, the display board can be inserted over the bolts; if fitted correctly, the display should protrude through the cut-outs. The display board is followed by four non-threaded spacers, and then by the control PCB. The shafts of the control potentiometers should protrude through specially-drilled holes in the display PCB and front panel. The boards are finally held in place by M3 securing nuts. The top right screw must be earthed using one of the tagged leads from the rear panel (see Figure 3.4). When this has been done, the three control knobs can be fitted. Finally, three connections must be made between the two boards; see Figure 1.16.

Take Note — Take Note — Take Note — Take Note

Disregard Figure 6.2 of the construction guide supplied with the kit, as this does not include earthing arrangements, and shows only a single-pole switch. A replacement wiring diagram is shown in Figure 1.19. Follow the revised circuit diagrams of Figures 1.13 and 1.14, rather than the versions printed in the manual.

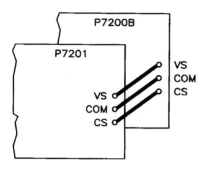

Figure 1.18   Inter-board wiring

For safety reasons, Figure 1.13 incorporates a double-pole mains switch. For clarity, Figure 1.19 shows only the mains wiring and that of the toroidal transformer's secondaries. Earthing arrangements and the reservoir capacitor wiring are shown in Figures 1.16 and 1.20 respectively. For the same reason, the display transformer's secondary wiring has been omitted from Figure 1.19; it is shown in Figure 1.21. Earthing apart, output terminals and the display transformer's secondary should be wired up first, as these may prove difficult to get at once the other wiring has been completed. At this stage, do not connect the fan, or the pass transistor PCBs (shown in Figure 1.22); this will be done during the testing stage. Connections at mains potential should be covered with heat-shrink sleeving, for safety's sake. The wires to the front-panel mains switch are attached via *Lucar* receptacles — don't forget those insulating boots! As seen in Figure 1.19, the primaries of both transformers are connected to the mains switch; extra care should be taken when connecting two wires to each terminal.

Power supply projects

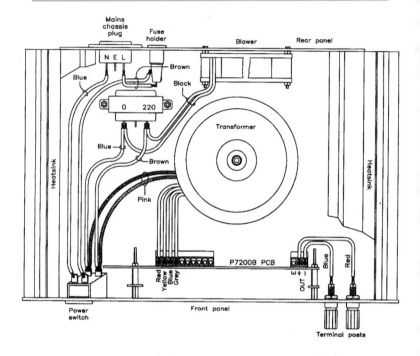

Figure 1.19 Wiring diagram — mains and PSU transformer secondary winding

Testing

Before testing, you should thoroughly check your work for any problems, such as short circuits, solder bridges, missing or misplaced components.

The following tests involve procedures to be carried out with the case top removed and 240 V a.c. mains connected. It is imperative that every possible precaution is taken to prevent electric shock. 240 V a.c. mains *can kill!*

42

Initial testing involves wiring a 240 V 40 W lamp bulb across the mains fuseholder; do not fit the fuse in the fuseholder at this stage. The bulb is now effectively wired in series with the transformer primary winding. If the lamp should illuminate brightly when mains voltage is applied, then too much current is being drawn by the power supply, which indicates a fault (for example, the

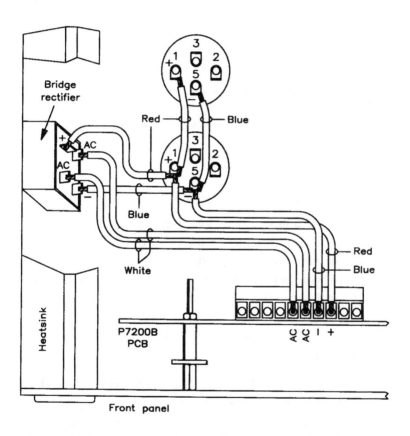

Figure 1.20   Connection of reservoir capacitors and bridge rectifier

Power supply projects

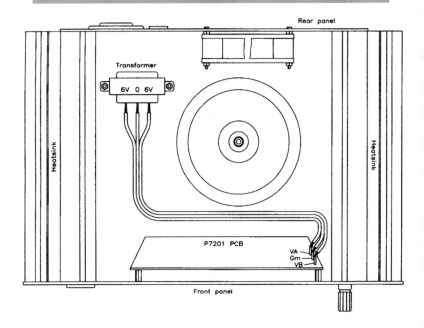

Figure 1.21   Display transformer

primary winding could be short-circuit). If the lamp is permanently unlit, there is likely to be an open-circuit on the primary winding of the transformer. If all is OK, the bulb will light for a short instant, and then go out again (or illuminate dimly) — as a surge of current flows through the transformer. If this does not happen, or one of the other symptoms is noted, then you should investigate the problem further.

With RV1/RV2 (main board) set to mid-positions, and the current limit potentiometer turned all the way to the right, testing can begin. When the power supply is switched on, the digital displays should now light up. A voltmeter connected between the output terminals

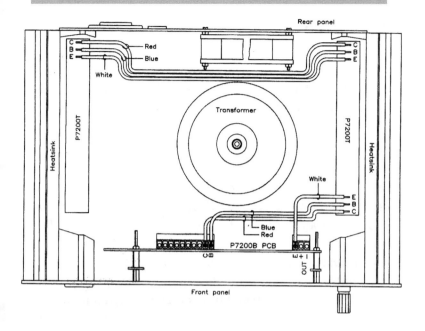

Figure 1.22 Wiring of pass transistor PCB

should give 0 to 30 V when the voltage adjustment control is advanced from left to right. At around 12 V, the relay should click, indicating that the other secondary winding of the unit's transformer has been switched into circuit.

Turn off the power (and unplug the unit!), then turn the display presets (RV1/RV2, display PCB) fully clockwise. Connect up the pass transistor PCBs, as shown in Figure 3.10. After switching the unit back on again, check that the voltage can be adjusted as before. Switch off and disconnect the mains supply; the lamp can now be disconnected, and a 4 A fuse fitted to the fuseholder. At this stage, the fan can be wired up — see Figure 3.7.

Power supply projects

Setting up

The following procedures should be carried out very carefully and not rushed. The performance of the project is dependant on the accuracy of these adjustments.

Power supply alignment

With the unit disconnected, turn the front panel current limiting control fully clockwise, and the two voltage controls (fine and coarse) to their midway positions. The two presets on the main panel need to be adjusted again; RV1 (maximum current adjustment) needs to be turned fully anti-clockwise, while RV2 (minimum current adjustment) needs to be turned fully clockwise. Alignment can now proceed; for this you will need an ammeter (i.e. multimeter set to measure current) with a range of at least 10 A. This should be connected across the power supply's output terminals. After powering up, the *current limit* LED will light up (the low resistance of the ammeter presents a virtual *dead short* to the PSU). RV1 should now be adjusted until the meter reads 10 A. Disconnect the meter, and turn the voltage/current limit adjustment controls fully anti-clockwise. RV2 should now be adjusted until the current limit LED lights up dimly.

Calibration of digital displays

After connecting a multimeter (set to 30 V range or higher) to the unit's output terminals, set the current limit control to 1 A. Switch on (both displays should read 000) and adjust the voltage controls so that the meter gives a reading of 30 V. Adjust the (voltage) *meter adjust* preset (RV1 on the display PCB, accessible from the front panel) until it reads the same as the multimeter. Turn the unit off, and set the multimeter to its 10 A (at least) current range. After switching the unit back on again, the current limit control should be set so that the multimeter reads 8 A. Adjust the (current) *meter adjust* preset, RV2 on the display panel, until the unit's display reads the same as the multimeter. Access to RV2 can also be gained from the front panel.

Your power supply is now ready for use!

Output voltage range:	0 to 30 V d.c.
Fine output control range:	1 V
Ripple:	0.5 mV r.m.s. (max)
Voltmeter resolution:	0.1 V
Output current:	0 to 8 A continuous (10 A peak)
Ammeter resolution:	0.01 A
Current limit indication:	LED
Power consumption:	300 W (max)
Dimensions:	330 mm (W) x 90 mm (H) x 215 mm (D)

Table 1.4   Specifications

10 A 30 V laboratory PSU parts list

Control PCB resistors

R1	6k8	1
R2	8k2	1
R3–8	4k7	6
R9–12	220 Ω	4
R13,14	2k2	2
R15,16	2k7	2
R17,18	820Ω	2
R19–22	22 k	4
R23,24	1k2	2
R25	100 k	1
R26	15 k	1
R27	100 Ω	1
R28,29	1 k	2
R30	220 k	1
R31	18 Ω	1
R32	270 k	1
R33	12 k	1
R34	27 k	1
R35	39 Ω	1
R36	180 Ω ($^1/_2$ W)	1
R37–40	constructed from resistance wire	1 m
RV1	100 Ω vertical preset	1
RV2	47 k vertical preset	1
RV3	4k7 pot log	1
RV4	10 k pot lin	1
RV5	1 k pot lin	1

Capacitors

C1	150 pF ceramic	1
C2	33 nF metallised polyester film	1
C3	68 nF metallised polyester film	1
C4	100 nF ceramic	1
C5	1 µF metallised polyester film	1
C6,7	10 µF 35 V electrolytic	2
C8,9	100 µF 35 V electrolytic	2
C10	470 µF 35 V electrolytic	1

Semiconductors

IC1	741	1
IC2,3	µA723	1
T1	BC557B (or equiv)	1
T2	BC547B (or equiv)	1
T3	BD646	1
D1–3	1N4000 series diode	3
ZD1	10 V 500 mW zener	1
ZD2	18 V 1.3 W zener	1
LD1	3 mm red LED	1

Miscellaneous

L1	4m7H	1
RY1	single pole changeover relay	1
J1–5	2-way PCB mounting terminal block	5
J6	3-way PCB mounting terminal block	1
	PCB pin	3

Power supply projects

14-pin DIL socket	2
8-pin DIL socket	1
PCB	1
heatsink for BD646	1
M3 12 mm screw	1
M3 nut	1
M3 shakeproof washer	1

Display PCB parts list

Resistors

R1–4	100 k	4
R5,6	22 k	2
R7,8	47 k	2
R9,10	470 Ω	2
RV1,2	1 k horizontal preset	2

Capacitors

C1,2	100 pF ceramic	2
C3–8	100 nF resin-dipped ceramic	6
C9,10	100 nF metallised polyester film	2
C11,12	220 nF metallised polyester film	2
C13,14	470 nF metallised polyester film	2
C15,16	1000 µF 16 V electrolytic	2

Semiconductors

IC1,2	7107	2
VR1	7805	1
VR2	7905	1
D1–4	1N4000 series diodes	4
DY1–6	7 segment display (common anode)	6

Miscellaneous

40-pin DIL socket	4
PCB pin	3
PCB	1
M3 12 mm screw	2
M3 nut	2
M3 shakeproof washer	2

Power supply projects

Pass transistor PCB parts list

Resistors

R41–45	220 mΩ 5 W	5

Semiconductors

TIP3055	5

Miscellaneous

$^1/_4$ in blade terminal	9
PCB pin	15
mica washer	5
insulating bush	5

Hardware parts list

Miscellaneous

B1	bridge rectifier	1
C11,12	4700 µF 63 V electrolytic capacitor	2
SW12	pole mains rocker switch	1
F1	20 mm panel mounting fuseholder	1
	4 A 20 mm fuse	1
	chassis mounting europlug	1
	europlug lead	1
TRANSFO	display transformer (6–0–6 V)	1
TRANSFO1	PSU main transformer (15–0–15 V 300 VA)	1
	mounting hardware (for above)	1
	cooling fan	1
	red terminal post	1
	black terminal post	1
	$^1/_4$ in blade receptacle	17
	$^1/_4$ in blade receptacle insulating boots	4
	heatsink	2
	front panel	1
	front panel foil	1
	control knob	3
	rear panel	1
	base	1
	top lid	1

Power supply projects

electrolytic capacitor mounting bracket	2
rubber feet	4
M3 12 mm zinc-plated bolt	12
M3 12 mm zinc-plated countersunk bolt	2
M3 15 mm hex head bolt	6
M4 25 mm zinc-plated bolt	1
M4 25 mm black bolt	4
M4 25 mm black countersunk bolt	4
M4 25 mm zinc-plated bolt	4
M4 45 mm zinc-plated bolt	4
M3 15 mm spacer	4
M3 solder tag	5
M3 washer	6
M3 shakeproof washer	20
M3 nut	28
M3 12 mm threaded bush	4

Optional (not in kit)

13 A nylon mains plug	1	(RW67X)
3 A fuse	1	(HQ32K)
insulating boot for fuseholder	1	(FT35Q)
insulating boot for chassis plug	1	(JK66W)
heat transfer compound	1	(FL79L)
BC lampholder	1	(FQ02C)
40 W mains lamp bulb	1	

The above items (excluding optional) are available as a kit, order as VF14Q

2 Chargers

Lead acid battery charger modules

Two types of lead-acid battery charger modules are available from Maplin Electronics; one of these (JY65V) is for charging 6 V batteries, while the other (JY64U) is suitable for 12 V types (see Table 2.1). Each charger features a regulated output suitable for charging sealed lead-acid batteries (rated at between 1 Ah and 8 Ah) at up to 500 mA, and has outputs for *power on* and *charge* indicators. The units feature current foldback and may be used for cyclic or trickle charging. The chargers operate from a low voltage a.c. supply and may be powered directly from a suitable transformer.

Figure 2.1 shows the input and output connections for the charger module; these are made via an 8-wire ribbon cable. Each charger requires an a.c. input within a specific range; this is between 11 and 13 V r.m.s. for the 6 V charger, or between 17 and 19 V r.m.s. for the 12 V version. It is important that the input voltage is kept within

Parameter	6 V charger	12 V charger
Input	a.c. 11 V to 13 V (15 VA)	a.c. 17 V to 19 V (20 VA)
Output current	700 mA	500 mA
DC maximum continuous load current	525 mA	375 mA
Operating temperature range	0 to 45°C	0 to 45°C
Charge indicator cut-off current	245 mA	175 mA
Applicable battery range dimensions	6 V 1 to 8 Ah	12 V 1 to 8 Ah
Dimensions	71 (Max 78) x 45 x 25 mm	71 (Max 78) x 45 x 25 mm

Table 2.1   Specifications of charger module

these limits, as far as possible, to maintain the correct output level. In addition, it is essential that the maximum input voltage rating is not exceeded as this could damage the unit. It should be noted that the charger is only suitable for charging batteries which have capacities between 1 Ah and 8 Ah — the unit is not suitable for use with batteries outside this range.

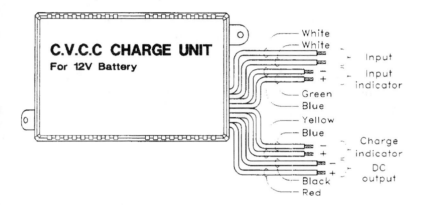

Figure 2.1   Module lead-out

Two battery charger kits available

Two different kits of parts are available (one for 6 V batteries, the other for 12 V batteries), each of which contains the basic hardware necessary to build a mains-powered lead-acid battery charger; as well as the module, the kit includes the necessary mains step-down transformer, case, fuse, indicators and so on.

Power supply projects

Construction

Because the module is pre-manufactured, very little construction is required — the unit needs only to be mounted in a box and wired up to the various inputs and outputs. Figure 2.2 shows how to assemble the charger, while Figure 2.3 shows the drilling details for the case.

The transformer is mounted in the box using two, M4 nuts, bolts and washers, and these should be tightened so that the transformer is completely stable and does not move around in the case. Figure 2.4 shows the wiring to the mains (primary) side of the transformer; for safe operation, this diagram *must* be followed exactly.

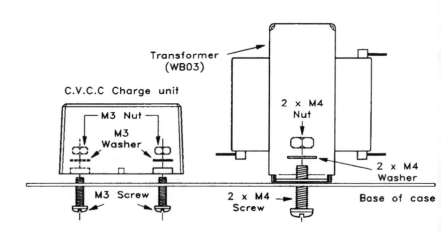

Figure 2.2 Assembly diagram

Chargers

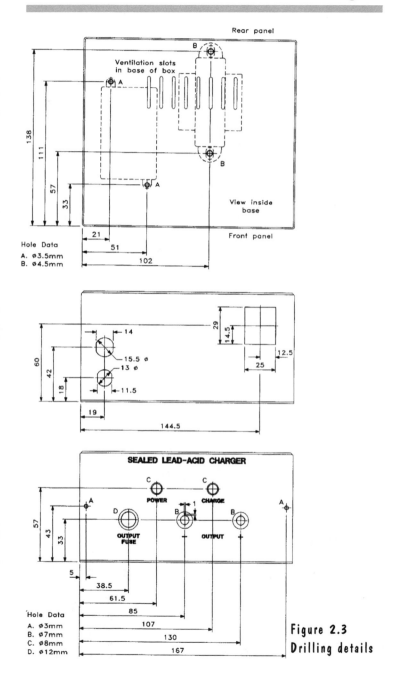

Rear panel

Ventilation slots
in base of box

B

A

111

138

57

33

View inside
base

21

51

Front panel

102

Hole Data
A. ⌀3.5mm
B. ⌀4.5mm

14

15.5 ⌀

13 ⌀

11.5

60

42

18

19

144.5

29

14.5

12.5

25

SEALED LEAD-ACID CHARGER

C POWER

C CHARGE

A

A

D OUTPUT FUSE

B

B OUTPUT

57

43

33

5

38.5

61.5

85

107

130

167

Hole Data
A. ⌀3mm
B. ⌀7mm
C. ⌀8mm
D. ⌀12mm

Figure 2.3
Drilling details

59

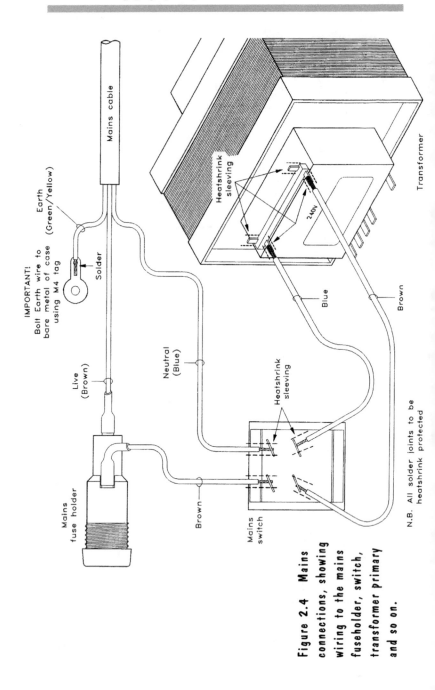

Figure 2.4 Mains connections, showing wiring to the mains fuseholder, switch, transformer primary and so on.

Mains cable

Earth (Green/Yellow)

IMPORTANT!
Bolt Earth wire to bare metal of case using M4 tag

Solder

Live (Brown)

Neutral (Blue)

Mains fuse holder

Brown

Mains switch

Heatshrink sleeving

Heatshrink sleeving

Blue

Brown

240v

Transformer

N.B. All solder joints to be heatshrink protected

The earth lead (green/yellow) is soldered to an M4 solder tag, which must be anchored to the case using an M4 nut, bolt and washer. The earthing tag must be bolted down as securely as possible — *this is an essential safety measure*. It is also important to make sure that all bare wires and terminals carrying live mains are adequately insulated; a length of heat-shrink sleeving is included in the kit for this purpose.

Also supplied in the kit is a strain-relief grommet, which should be clipped around the mains lead and inserted into a cable entry hole drilled in the back panel. This will prevent the mains lead from coming into contact with the metal case, and stop the cable from being tugged out of the case accidentally. The grommet is quite a tight fit and may require some effort in pushing it into position.

Figures 2.5 and 2.6 show the wiring from the secondary side of the transformer to the charger module; note that the wiring of the 6 V unit is different to that of the 12 V unit. The battery charger wires are colour-coded for identification purposes. The charger output is wired to binding posts on the front panel; the red lead from the charger module is wired to the red binding post while the black wire goes to the black binding post. The power-on indicator LED (red) and charge indicator LED (green) are both mounted on the front panel. Both are held in place using the LED clips supplied in the kit.

The unit has both a primary and a secondary fuse. The $1^{1}/_{4}$ in mains fuseholder is mounted on the back panel of the case and should be fitted with a 100 mA fuse. The secondary fuseholder, mounted on the front panel of the case, is a 20 mm type, and should be fitted with an 800 mA

Power supply projects

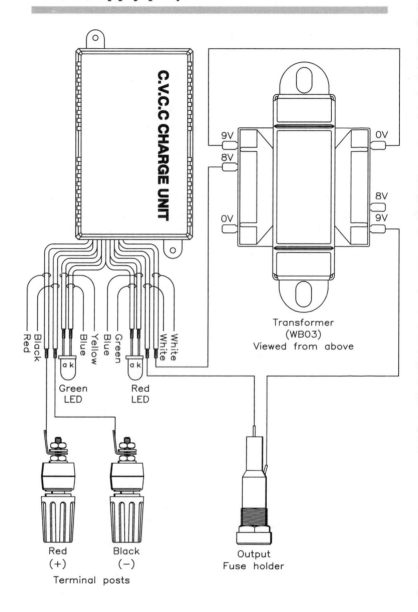

Figure 2.5   6 V charger — low voltage wiring information

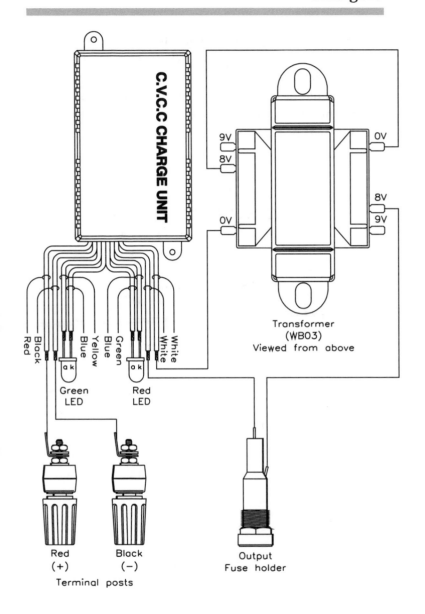

Figure 2.6   12 V charger — low voltage wiring information

Power supply projects

fuse. One of each type of fuse is supplied in the kit. Note that the charger should always be disconnected from the mains before changing the fuses. Figure 2.7 shows a typical front panel layout, while Figure 2.8 shows how to mount the fuseholders. Figure 2.9 shows how to mount the terminal posts.

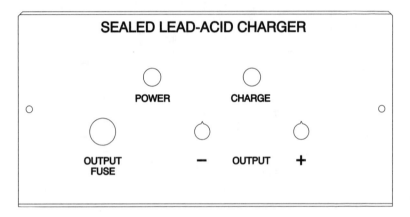

Figure 2.7   Suggested front panel layout

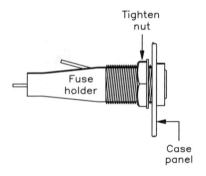

Figure 2.8   Mounting the fuseholders

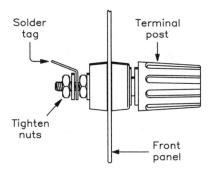

Solder
tag

Terminal
post

Tighten
nuts

Front
panel

Figure 2.9   Mounting the terminal posts

Testing

Before applying power to the unit, double-check your work to make sure that everything is in place and wired correctly. Make sure the box lid is in place before testing the unit. Always disconnect the mains supply before attempting any work on the charger or removing the case lid — remember that *mains voltage can kill.*

Connect the charger to the mains supply and switch on the unit via power switch S1, located at the rear of the case. The red *power on* LED on the front of the unit should illuminate, confirming that the unit is powered up. Temporarily connect a piece of wire between the charger output terminals (TB1 and TB2) to act as a shorting link. During the period that the wire is connected, the *charge* LED should illuminate. If all is well, the shorting link should be disconnected — the charger is now ready for use.

Power supply projects

When connecting a lead-acid battery to the charger, it is important that the connections are of the correct polarity; the red binding post is connected to the positive (+) side of the battery, while the black binding post is connected to the negative (−) side of the battery. To test the unit fully, a multimeter, set to read current, may be connected between the red terminal on the charger and the positive (+) terminal of the battery; a direct connection is made between the negative (−) terminal of the battery and the black terminal on the charger. The charge current can then be monitored and this should be around 500 mA initially (up to 700 mA for the 6 V charger) for a fully discharged battery, or between 0 and 10 mA for a charged battery. When a new battery is charged up for

12 V version

Supply voltage	Mains (240 V a.c. r.m.s.)
Output current	525 mA maximum
Charge LED cut-off point	175 mA
Suitable 12 V battery capacity	1 to 8 Ah

6 V version

Supply voltage	Mains (240 V a.c. r.m.s.)
Output current	700 mA maximum
Charge LED cut-off point	175 mA
Suitable 6 V battery capacity	1 to 8 Ah

Please note: never apply 240 V mains directly to the input of the charger module — or severe damage will result. Always use a suitable step-down transformer.

Table 2.2 Specifications of prototype finished chargers

the first time, this rule may not be followed, the charge current fluctuating in an unpredictable manner. The current values stated are only approximate and may vary considerably, depending on the type and condition of the battery. *Please note*: *never* connect a multimeter, set to read current, directly between the negative and positive terminals of a lead-acid battery, as this may cause irreparable damage to the meter. The charge LED is only intended to give an approximate guide to the charge condition, indicating when the charge current is above a preset threshold (see Table 2.1). The unit will continue to charge the battery below this threshold at a lower current, and in this case the charge LED will not be illuminated.

Using the charger

The charger is suitable for use with 6 V and 12 V sealed lead-acid batteries with capacities of between 1 Ah and 8 Ah. It is not suitable for use with batteries outside this range. When not in use, the charger should be positioned in a well-ventilated area allowing free air-flow around the unit. The ventilation holes in the box should not be obstructed as this may result in overheating. Although reasonable care has been taken to ensure that the charger is safe, it is recommended that the unit is not left unsupervised for long periods during the charging process. The mains supply should, of course, always be disconnected when the charger is not in use.

6 V sealed lead-acid battery charger parts list

6 V SLA charger	1	(JY65V)
$1^1/_2$ A 9 V Tr	1	(WB03D)
1608 steel case	1	(XJ28F)
2 mA 5 mm LED red	1	(UK48C)
2 mA 5 mm LED green	1	(UK49D)
$1^1/_4$ in clickcatch F/H	1	(FA39N)
large term post black	1	(HF02C)
large term post red	1	(HF07H)
5R2 SR grommet	1	(LR48C)
DPST rocker	1	(YR69A)
M4 isoshake	1	(BF43W)
M3 isoshake	1	(BF44X)
M4 x 20 mm pozi screw	1	(JC75S)
M3 x 20 mm pozi screw	1	(JC71N)
M3 steel nut	1	(JD61R)
M4 steel nut	1	(JD60Q)
CP32 heat shrink	1	(BF88V)
20 flush fuseholder	1	(KU33L)
20 mm 800 mA fuse	1	(RA03D)
$1^1/_4$ A/S 100 mA fuse	1	(UK58N)
5 mm LED clip convex	1	(UK14Q)
min mains black	2	(XR01B)
M4 isotag	1	(LR63T)
10 m 16/0.2 wire black	1	(FA26D)
CP64 heat shrink	1	(BF90X)
instruction leaflet	1	(XT73Q)
constructors' guide	1	(XH79L)

12 V sealed lead-acid battery charger parts list

12 V SLA charger	1	(JY64U)
$1^1/_2$ A 9 V Tr	1	(WB03D)
1608 steel case	1	(XJ28F)
2 mA 5 mm LED red	1	(UK48C)
2 mA 5 mm LED green	1	(UK49D)
5 mm LED clip convex	2	(UK14Q)
min mains black	2	(XR01B)
$1^1/_4$ in clickcatch F/H	1	(FA39N)
large term post black	1	(HF02C)
large term post red	1	(HF07H)
5R2 SR grommet	1	(LR48C)
DPST rocker	1	(YR69A)
CP 32 heat shrink	1	(BF88V)
20 flush fuseholder	1	(KU33L)
800 mA 20 mm fuse	1	(RA03D)
100 mA $1^1/_4$ A/S fuse	1	(UK58N)
M4 isoshake	1	(BF43W)
M3 isoshake	1	(BF44X)
20 mm M4 pozi screw	1	(JC75S)
20 mm M3 pozi screw	1	(JC71N)
M3 steel nut	1	(JD61R)
M4 steel nut	1	(JD60Q)
M4 isotag	1	(LR63T)
10 m 16/0.2 wire black	1	(FA26D)
CP 64 heat shrink	1	(BF90X)
instruction leaflet	1	(XT73Q)
constructors' guide	1	(XH79L)

Power supply projects

Optional (not in kit)

10 m 16/0.2 wire red	1	(FA33L)
10 m 16/0.2 wire black	1	(FA26D)
4 mm croc clip black	1	(HF23A)
4 mm croc clip red	1	(HF24B)
charger clip	2	(HF26D)
large battery clip red	1	(FS86T)
large battery clip black	1	(FS87U)
13 A nylon plug	1	(RW67X)
2 A fuse plug	1	(HQ31J)

The above items (excluding optionals) are available as kits, order as LP91Y (6 V Lead Acid Charger) and LP73Q (12 V Lead Acid Charger)

70

Slow charger for Ni-Cads

High performance model racing cars and electric powered model aircraft make huge demands on their Ni-Cad
battery racing packs. For optimum model performance,
Ni-Cads must be able to deliver extremely high currents
while maintaining the rated terminal voltages for long
periods of use. Because regular rapid charging techniques are employed out in the field, Ni-Cad packs tend
to *memorise* or suffer from reduced capacity, thus pre-
venting the full charge/discharge parameters from being
reached. Early warning signs of this effect become ap-
parent when running models; racing cars tend to become
less lively and top-lap speeds drop off, aircraft lack
height or exhibit a reduced rate of climb and perhaps
more noticeably, the model's running time becomes in-
creasingly shorter. Under rapid charging conditions,
older battery packs may heat up after just a few minutes
whereas previously they remained cool — although this
could also signify excessive abuse or cell breakdown!

To maintain maximum Ni-Cad capacity, it is necessary
to regularly slow charge/discharge the pack at regular
intervals and for reduced capacity Ni-Cads, a sequential
cycle of slow charge/discharging over several days can
restore much of the original capacity.

How it works

The Slow Charger is a very simple project to build and is
based on the well known constant current principle. Both
7.2 V and 8.4 V Ni-Cad packs can be used on this system

71

Power supply projects

and charged at 120 mA (±5 mA) which is safe for most packs available with charging periods up to 15 hours. In fact Maplin racing packs may be overcharged up to 20,000 hours without problem on this system! With reference to Figure 2.10, the charger requires a separate supply of 15 V d.c. provided by the *unregulated* mains adapter type XX09K, and this connection is reverse polarity protected by full wave bridge rectifier, BR1. The battery pack connected between pins 3 and 4 is prevented from discharging back through the charger by diode D1. With the battery out of circuit, base current via R1 is 12 mA (15 V supply) and allowing for 0.62 V_{be} and an emitter load of 11 Ω, the voltage drop across LD1 is 0.75 V; the LED requires approximately 2 V to conduct, therefore D1 does not illuminate. With a battery connected in the collector of TR1, current flows from the bridge through R2/R3 parallel pair. A voltage drop of 1.35 V appears across these resistors at TR1 emitter and is clamped 0.65 V higher at TR1 base by the LED. As the voltage across LD1 is now 1.35 + 0.65 = 2 V, the LED conducts and collector current through the Ni-Cad is defined as 1.35 volts/11 Ω = 123 mA. Capacitor C1 reduces any ripple present from the external PSU, thus maintaining the d.c. constant current characteristics of the charger.

Construction

Figure 2.11 shows the very simple layout of the circuit board. Refer to the parts list and constructors' guide for assembly techniques if in doubt and ensure that the (+) symbol on BR1 aligns with the (+) symbol on the PCB. TR1 must be fitted with the metal case flat down onto

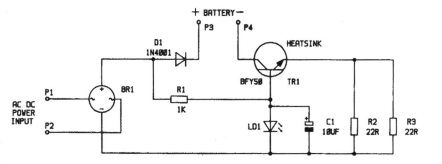

Figure 2.10 Circuit

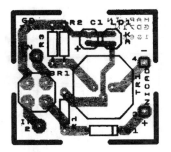

Figure 2.11 PCB track and layout

the PCB and the clip-on heatsink attached. When fitting LD1, place the leads so they are just protruding through their track pads and solder in place. This will allow the LED to stand approximately 18 mm above the PCB, leaving enough room for fitting into the box (Figure 2.12). Solder all components and clip off excess wire ends. Insert four pins from the track side and push their heads down onto the PCB with a soldering iron prior to soldering.

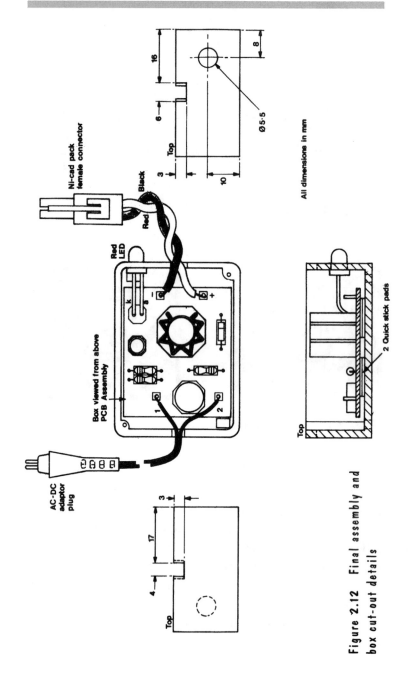

Figure 2.12 Final assembly and box cut-out details

Assembly

The completed module is fitted into a plastic box and held in place with double sided stick pads as shown in Figure 5.3. A 5 mm hole will need to be drilled in one end of the box for LD1 and slots filed out at both ends to take the PSU and battery cables. Connect the adapter plug wires to input pins 1 and 2 — do not worry which way around — and solder the female connector positive (red) lead to pin 3 and the negative (black) lead to pin 4. Mount two quickstick pads at each end of the PCB, on the track side only and place another row of pads on top of them so that they are two high. Remove any remaining paper strips and place the module into the box, LD1 end first, inserting the LED through the previously drilled hole in the box end. Twist the module round into the box and push firmly down so the pads have a good purchase on the box bottom. If the end panel slots have been filed correctly then the cables should fit into them without standing proud, otherwise the box lid will not be a flush fit.

Using the charger

The most common Ni-Cad racing packs used in models are either 1.2 Ah, 1.4 Ah or 1.8 Ah versions. At a charging rate of 0.12 Ah it will take 10 to 15 hours for these batteries to reach full capacity. Always check the manufacturers' recommendations for charging first in case variations are encountered. Of course it is almost

Power supply projects

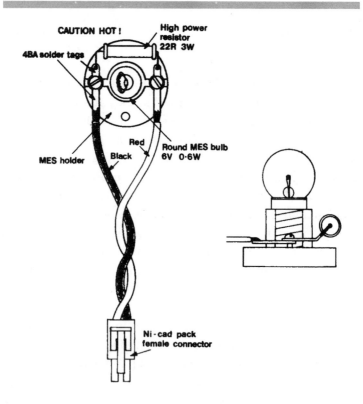

Figure 2.13   Fixed load discharger

5.4

impossible to know the state of charge of a battery at any one time, so that charge times can be calculated, therefore it is advisable to discharge the battery first. The diagram in Figure 5.4 shows a 6 V, 0.6 W (100 mA) torch type lamp connected in parallel with a 22 Ω resistor. For 1.2 Ah packs the discharge rate will average 0.425 A, falling as the Ni-Cads near discharge, and the lamp will gradually become dimmer. It will take about 3

Chargers

hours to discharge a racing pack and power dissipated by the resistor is approximately 2.5 W, which means it becomes very hot to the touch! 8.4 V and 1.8 Ah packs will require just over 4 hours to discharge with this system. Always slow charge the racing pack before use whenever possible as doing so will ensure the maximum charge capacity being available. After a heavy session of rapid charging, allow the battery to slow discharge into the fixed load shown and then slow charge for 15 hours. Repeat this for two or three cycles over a few days to keep the battery in good working order.

77

Power supply projects

Slow charger parts list

Resistors — All 0.6 W 1% metal film

R1	1 k	1	(M1K)
R2,3	22 Ω	2	(M22R)

Capacitors

C1	10 µF 16 V minelect	1 (YY34M)

Semiconductors

D1	1N4001	1 (QL73Q)
LD1	LED red	1 (WL27E)
TR1	BFY50	1 (QF27E)
BR1	W005	1 (QL37S)

Miscellaneous

PC board	1 (GD71N)
veropin 2141	1 (FL21X)
quickstick pads	1 (HB22Y)
heatsink	1 (FL78K)
box 1521	1 (FK72P)
female race pack lead	1 (JG05F)
d.c. power mains adaptor unreg	1 (XX09K)

Optional

6 V 0.6 W MES bulb	1 (WL78K)
MES batten holder	1 (RX86T)
4BA tag	1 (BF28F)
female race pack lead	1 (JG05F)
22 Ω 3 W high power resistor	1 (W22R)

A kit of the above parts (excluding optional) is available, order as LM39N

Rapid charger for Ni-Cads

Build this Ni-Cad Rapid Charger and put real power into your radio control model car. The unit is powered from a conventional 12 volt lead acid car battery, which can be left in your vehicle or removed for track side use. Housed in a tough steel case the Rapid Charger is ideally suited for use at outdoor offroad race meetings.

Introduction

The sport of competitive off-road model car racing has become very popular over the past few years. The success of this hobby is mainly due to the increasing technical sophistication of the models. Four wheel independent suspension and four wheel drive cars have now become commonplace. The majority of models use small yet powerful electric motors in preference to the model internal combustion engine.

These electric motors, when in a race; draw several amps of current from the battery, rapidly draining the power from the cells. With present battery technology the rechargeable nickel cadmium cell is most suited for this application. There are two main configurations of cells used at present, the 6-cell providing 7.2 volts and the 7-cell giving 8.4 volts. The physical arrangement of cells used in any particular model could be a flat, hump or tunnel pack. All these racing packs have two short lengths of high current silicone rubber insulated wire, terminated with a non-reversible male power connector.

Power supply projects

The normal charge rate for the 1.2 Ah NiCad cells used on the prototype was 120 mA for 16 hours, with a continuous overcharge period of more than 20,000 hours. However, an accelerated charge of 480 mA for 3.5 hours can be used in complete safety, with a continuous overcharge period of more than 10 days. When rapid charging at currents in excess of 1 amp for 15 to 30 minutes, care must be taken not to overcharge the cells as damage will occur. It is for this reason an electronic timer is used to shut off the high current at the end of the selected period and put a trickle charge on the cells.

Circuit description

Referring to Figure 2.14, the positive d.c. voltage from the lead acid car battery is first applied to FS1, the 5 A, 20 mm anti-surge fuse. This protects the circuit from burning out if a faulty short circuit racing pack is connected to the unit. The d.c. supply entering the circuit must have the correct polarity, otherwise damage will occur to the semiconductors and polarised components. To prevent this, a diode, D2, has to have the positive supply voltage applied to its anode before the d.c. power can pass to the relay control and timer circuits. D1 is a high current diode which prevents the Ni-Cad racing pack, B1, from receiving a reverse polarity charge or discharging back into the circuit if the supply is removed.

The timer circuit comprises CMOS integrated circuits IC1, a 4020BE, and IC2, a 4060BE. IC1 is a 14-stage ripple counter whereas IC2 is a 14-stage ripple counter and oscillator. It is the frequency and stability of this oscil-

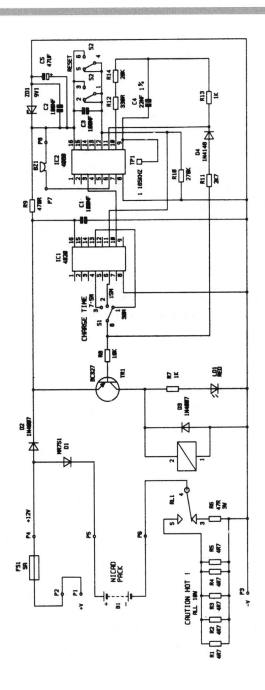

Figure 2.14 Circuit

Power supply projects

lator that will determine the accuracy of the selected charge period. There are two main influences on oscillator stability, supply voltage and ambient temperature. To minimise the effect of supply voltage fluctuations a 9.1 V zener diode, ZD1, limits the voltage fed to pin 16 of IC2. This supply is then decoupled by capacitors C2 and C4 to remove any electrical noise. To maintain frequency accuracy over a range of temperatures, high stability components are used in the oscillator circuit. The frequency of which is set by the values of C5 a 1% close tolerance polystyrene capacitor and 1% resistors R12, R14.

To obtain the desired charging period times of 7.5, 15 and 30 minutes the oscillator must run at a frequency of 1.165 kHz (858.3 µs). This frequency may vary slightly due to component tolerance and ambient temperature, it can be measured using a frequency counter at TP1. The output from the oscillator stage of IC2 is then divided by its binary counters to produce two much longer time periods, one of 54.93 ms at pin 6 and 7.031 seconds at pin 3. The output on pin 6 is used to drive the piezo sounder BZ1. This produces an authentic ticking clock sound while the Ni-Cad pack is charging and stops at the end of the charge period. The output on pin 3 is connected to pin 10, the clock input of IC1 for further division.

The full supply voltage is connected to pin 16 of 1C1 and is decoupled by C1 to remove any electrical noise. The three outputs used for the relay control circuit are pin 4 the divide by 64, pin 6 the divide by 128 and pin 13 the divide by 256 outputs. This corresponds to the 7.5, 15 and 30 minute time periods. The desired period having been selected by S1 then feeds the normally low signal to R8 in the base of TR1 and through R11 and D4 to the oscillator stage of IC2. When the signal goes high at the end of the time period D4 is forward biased and pulls up pin 11 of IC2 stopping the oscillator. The system can be set going again by pressing the reset switch, S2. This takes pin 11 of IC1 and pin 12 of IC2 high, thus resetting their binary counters.

While the timer is running, TR1 is biased on and current will flow in its collector circuit. This results in the relay, RL1, becoming energised and the red LED indicator, LD1, to light. RL1 is used to select the full charge current or the much lower trickle charge for B1. When energised RL1 selects the resistor network comprising of five 4.7 Ω ten watt resistors, R1 to R5. The total resistance of the network is 0.94 Ω and the power dissipation capacity is fifty watts. This high power dissipation is necessary when high current rapid charging is occurring. When not energised RL1 selects R6 the 47 Ω three watt resistor, allowing less than 100 mA to flow into the Ni-Cad pack. TR1 is biased off at the end of the selected time period by the voltage applied via R8 to its base. The diode, D3, across the coil in RL1 is there to suppress the high voltage pulse which is generated when the current stops flowing.

PCB assembly

The PCB is a single-sided fibre glass type, chosen for maximum durability and heat resistance. Removal of a misplaced component is quite difficult, so please double-check each component type, value and its polarity where appropriate, before soldering! For further information on component identification and soldering techniques please refer to the constructors' guide included in the kit.

The PCB has a printed legend to assist you in correctly positioning each item, see Figure 2.15. The sequence in which the components are fitted is not critical. However, it is easier to start with the smaller components. Begin with the metal film 0.6 W resistors, then mount the disc ceramic capacitors, C1, C2, C3 and C4 the close tolerance polystyrene capacitor. The polarity of the electrolytic capacitor, C5, is shown by a plus sign (+) matching that on the PCB legend. However on some capacitors the polarity is designated by a negative symbol (–) in which case the lead nearest this symbol goes away from the positive sign on the legend.

When fitting the transistor, TR1, you must carefully match the case to the outline shown. The diodes, D2, D3, D4 and ZD3, have a band at one end to identify the cathode connection. The high current diode, D1, has a ring at one end to identify its cathode. Be sure to position them accordingly.

Next, install the slide switch, S1, making certain that it is pushed down firmly on to the surface of the PCB. Before fitting the reset switch, S2, you must first convert it from locking to momentary push non-locking operation. A special nylon retainer clip is supplied with the switch, which replaces the wire retainer, converting it to momentary action. With either removed the plunger will be

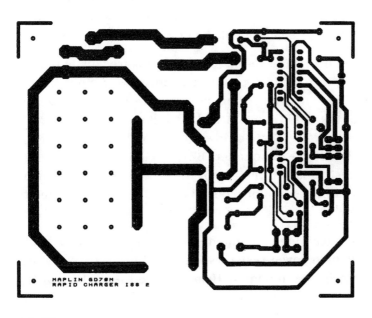

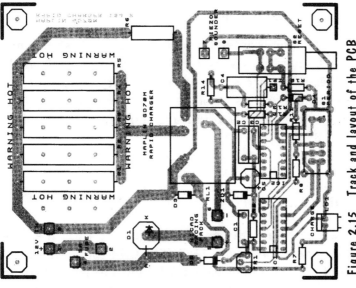

Figure 2.15 Track and layout of the PCB

Power supply projects

forced out by the spring, so keep it firmly held in. When fitting the switch make certain that it is pushed down firmly on the surface of the PCB.

When fitting the 16-pin IC sockets ensure you match the notch with the block on the legend. Now carefully install, IC1, and, IC2, into their appropriate sockets. Next, install the red PCB mounted LED and relay making certain that they are pushed down firmly on the surface of the PCB. The piezo sounder, BZ1, is mounted using a self-adhesive pad to the top of the relay. The sounder may have different coloured leads but either can be taken to P7 or P8.

The remaining components to be fitted are the 3 W and 10 W high power resistors. The five 10 W, 4.7 Ω resistors, R1 to R5 are mounted 10 mm above the surface of the PCB, over the ventilation holes, see Figure 2.16. Finally install R6, the 3 W, 47 Ω resistor making certain that it is also mounted 10 mm above the PCB.

This completes assembly of the PCB and you should now check your work very carefully, ensuring that all the solder joints are sound. It is also very important that the bottom, track side of the circuit board does not have any trimmed component leads standing proud by more than 3 mm.

Final assembly

The case which the unit is designed to fit is the *Steel Instrument Case type 1105* (XJ25C). Remove the black painted top from the case and follow the drilling instructions in Figure 2.17 when preparing the base. It has to

86

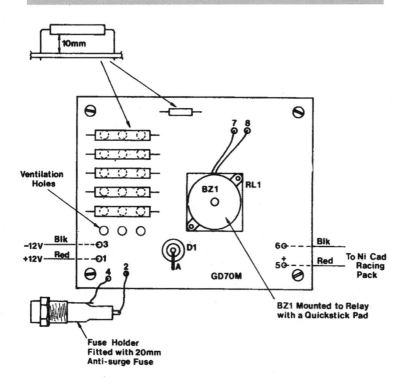

Figure 2.16   Component mounting on the PCB

have a number of holes drilled into it for ventilation and PCB fixing. The top cover has ventilation holes already punched and this can be used as a guide for checking the positioning of the base ventilation holes.

Remove the front and back panels from the case and follow the drilling instructions in Figure 2.18 when preparing them. The back panel has two holes drilled into it, one for the fuseholder and the other for the grommet at the power input. The self-adhesive front trim can be used as a guide for checking the positioning of the

87

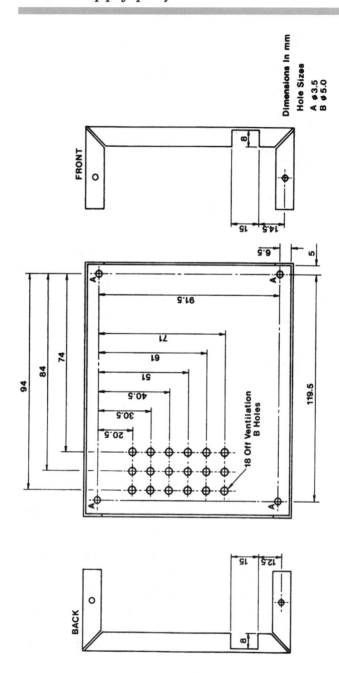

Figure 2.17 Case drilling

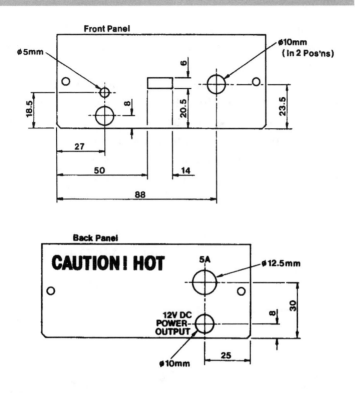

Figure 2.18 Front and rear panel drilling

holes in the front panel. Having drilled the holes at the
same time clearing them of any swarf, clean the front
panel and remove the protective backing from the self-
adhesive front trim. Carefully position and firmly push
down using a dry, clean cloth until it is securely in place.
Install the small grommets into the panels as shown in
Figure 2.19(a). Then using the self-tapping screws supplied with the case refit the front and back panels. Install
the four threaded spacers at the PCB fixing holes using
M3 bolts.

Power supply projects

Next prepare two 30 mm lengths of red high current wire. Remove 5 mm of insulation from each and solder them to the PCB at P2 and P4, see Figure 2.16. Note that they are inserted from the component side and the other ends are connected to the fuse holder at a later stage. The two, metre long, power input cables have the appropriate large insulated battery clip fitted at one end and 7 mm of insulation removed from the other, see Figure 5.6(b). The female Ni-Cad power connector is supplied with a 150 mm length of red and black high current wire already fitted. Ensure that 7 mm of insulation is removed from the free ends of the cable. Each power cable is then fed through its appropriate grommet into the case, see Figure 2.19(a).

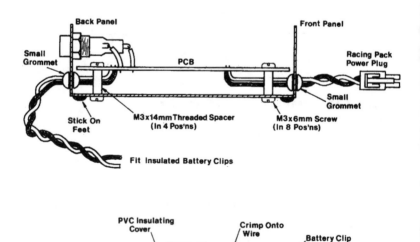

Figure 2.19 (a) Mounting assembly in case, (b) fitting battery clips

In the next stage the high current cables are inserted into the PCB from the copper track side. The red power input cable goes to P1, and the black to P3. The red Ni-Cad power cable goes to P5, and the black to P6, see Figure 2.16. Next secure the PCB assembly onto the threaded spacers using four M3 bolts. Install the 20 mm fuseholder onto the back panel and solder its terminals to the red wires from P2, P4. Finally, fit the 5 A 20 mm antisurge fuse and push the black round button onto the reset switch, S2. Do not fit the black case top until the testing stage is successfully completed.

Testing the unit

All the tests can be made with a minimum of equipment. You will need an electronic digital, or analogue, moving coil multimeter, preferably with a 10 A d.c. current range. The power source can be a 12 V lead acid car battery, or a 12 V to 14 V d.c. high power regulated supply, capable of up to at least 5 A. To check the timing accuracy of the unit you will require a watch or clock.

The following test results were obtained from the proto-type using a digital multimeter and a 12 V d.c. power supply. Two 1200 mAh Ni-Cad racing packs were used for the charging tests, a 6-cell 7.2 V and a 7-cell 8.4 V. Note that before a racing pack can be rapid charged *it must first be in its discharged state*, of less than 1 V per cell.

The first test is to ensure that there are no short circuits before you connect the supply. Set your multimeter to read *ohms* on its resistance range and connect the

Power supply projects

probes to the large battery clips on the power input cable. The reading obtained with the probes either way round should be greater than 1000 Ω. This test procedure is then applied to the terminals in the female power output connector and should give similar readings.

Next position the charge time switch S1, to its 7.5 minute setting and connect the large black battery clip to the negative d.c. supply. *Do not* fit a Ni-Cad racing pack to the female power connector at this time! To monitor the supply current set your meter to read mA and place it in the positive line to the large red battery clip. The unit may start up on its own, disregard this and push the reset switch, S2. Start timing the unit with your watch or clock. When the unit is in its full charge mode the red LED, LD1, should be illuminated and the piezo sounder, BZ1, should produce a clock like ticking sound. The current reading should be approximately 50 mA. At the end of the full charge period the LED will go out and the sounder should stop ticking. The current reading should drop to approximately 5 mA. Repeat this procedure for the 15 and 30 minute settings of the time switch and when successfully completed proceed to the final testing stage.

Final testing

Set your meter to its 10 A d.c. range. If a high current range of more than 5 A is not available, then remove the meter from the positive line and connect the large red battery clip, directly to the supply. Using the 12 V information provided in Table 2.3, set the time switch for 15 minutes for a 7.2 V Ni-Cad pack and 30 minutes for an 8.4 V pack. Plug the Ni-Cad pack on to the female connector and press the reset button. *Warning!* The 10 W high power resistors, R1 to R5, will become very hot during the full charge period. The currents shown in Table 2.3 and Figure 2.20(a) and 2.20(b) are dependent upon the individual condition of the racing pack being charged. At the end of the full charge period the unit will automatically switch over to a trickle charge of approximately 70 mA for a 7.2 V pack and 40 mA for a 8.4 V pack. This completes the testing of the rapid charger. Finally secure the case top with the four self-tapping screws. The rapid charger is now ready for use.

Using the rapid charger

When using the charger with a lead acid battery in your car, it is possible to increase the voltage supplied to the unit. If the engine in your vehicle is left running, the voltage across the terminals will increase as the alternator charges the battery. You must follow the charging times given in Table 2.3 for the different supply voltages. In addition to this the following operating procedure should be observed:

Supply voltage:	12 V lead-acid car battery
Supply current:	5 A maximum
Batteries charged:	6-cell 7.2 V and 7-cell 8.4 V racing packs
Charge time:	7.5, 15 and 30 minutes
Charge current:	3 A for 7.2 V packs and 1.8 A for 8.4 V packs
Trickle charge:	60 mA for 7.2 V packs and 40 mA for 8.4 V packs
Audio/visual:	Red LED charging indicator
	Piezo ticker sounder
Case dimensions:	Width 118 mm, length 143 mm, height 51 mm

Table 2.3   Specification of prototype

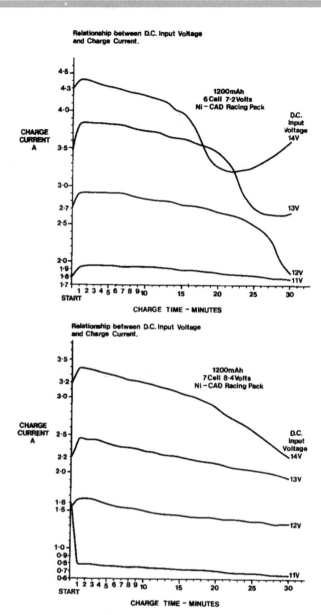

Figure 2.20 (a) 6 cell current charging graph (b) 7 cell current charging graph

● connect the large black battery clip to the negative terminal of the 12 V power source,

● connect the large red battery clip to the positive terminal of the 12 V power source,

● select the necessary charging time for the type of Ni-Cad pack and supply voltage, see Table 5.1,

● connect the Ni-Cad pack to the female power connector,

● press the reset button,

● at the end of the charge period the Ni-Cad pack can be removed when required.

— Take Note — Take Note — Take Note — Take Note —

Warning!

Do not attempt to charge a *hot* Ni-Cad pack.

Do not attempt to charge a Ni-Cad pack unless in its discharged state.

Do not over charge a Ni-Cad pack.

Do not obstruct the ventilation holes in the case of the rapid charger.

Power supply projects

Operating tips

Always carry some spare 20 mm 5 A antisurge fuses. Use a digital, or analogue, moving coil meter to measure the supply voltage. Have one Ni-Cad pack charging and another ready for use. Occasionally use a slow charger to maintain the condition of the Ni-Cad pack.

Rapid charger parts list

Resistors — All 0.6 W 1% metal film unless stated

R1–5	4Ω7 10 W WW	5	(H4R7)
R6	47 Ω 3 W WW	1	(W47R)
R7,13	1 k	2	(M1K)
R8	10 k	1	(M10K)
R9	470 Ω	1	(M470R)
R10	270 k	1	(M270K)
R11	2k7	1	(M2K7)
R12	330 Ω	1	(M330R)
R14	20 k	1	(M20K)

Capacitors

C1,2,3	100 nF disc ceramic	3	(BX03D)
C4	47 µF 16 V min elec	1	(YY37S)
C5	22 nF 1% polystyrene	1	(BX87U)

Semiconductors

IC1	4020BE	1	(QX11M)
IC2	4060BE	1	(QW40T)
D1	MR751	1	(YH96E)
D2,3	1N4007	2	(QL79L)
D4	1N4148	1	(QL80B)
LD1	LED red	1	(QY86T)
ZD1	BZY88C9V1	1	(QH13P)
TR1	BC327	1	(QB66W)

Power supply projects

Miscellaneous

S1	2-way momentary switch	1	(FH67X)
S2	2-pole 3 position R/A switch	1	(FV02C)
	20 mm safuseholder	1	(RX96E)
	5 A 20 mm anti-surge fuse	1	(RA12N)
	grommet small	2	(FW59P)
RL1	8 A 12 V flat relay	1	(HY20W)
	16-pin DIL socket	2	(BL19V)
	wire red	2	(XR59P)
	wire black	2	(XR57M)
	piezo sounder	1	(FM59P)
	female race pack lead	1	(JG05F)
	PC board	1	(GD70M)
	insulated battery charger clip red	1	(FS86T)
	insulated battery charger clip black	1	(FS87U)

Optional (not in kit)

M3 6 mm bolt	1	(BF51F)
M3 14 mm threaded spacer	1	(FG38R)
steel instrument case model 1105	1	(XJ25C)
quickstick pads	1	(HB22Y)
front trim	1	(JG19V)
round latchbutton black	1	(FL31J)

The above parts (excluding Optional) are available as a kit, order as LM40T

3 Inverters

12 V/230 V inverter

This device enables mains operated equipment, with a power consumption of 250 W or less, to be powered from a high current +12 V d.c. power source. Power sources include a car battery or alternator. This could be very useful if you need 230 V a.c. mains voltage in the event of a power cut, for example, to power an electric pump and timer circuit for gas central heating, or domestic electric lighting. It could even be used as a backup supply to protect sensitive computer equipment in the event of a power cut. Other uses may be to supply 230 V a.c. in areas where it would not normally be present, such as in a car, caravan, outhouse, or even on a boat.

Circuit description

A block diagram of the inverter is shown in Figure 3.1, and the *quasi* sine-wave output is shown in Figure 3.2. Referring to Figure 3.3 for the circuit diagram, note that reverse polarity protection of the circuit is provided by the fuse F1. In the event of wrong polarity connection, diode D17 will conduct and blow the fuse. This method of polarity protection is preferred to a series connected diode (as there is no voltage drop across the diode). This is important when the criteria for efficiency and energy is at a premium, especially when batteries are being used. Capacitor C18 provides the main decoupling in the circuit.

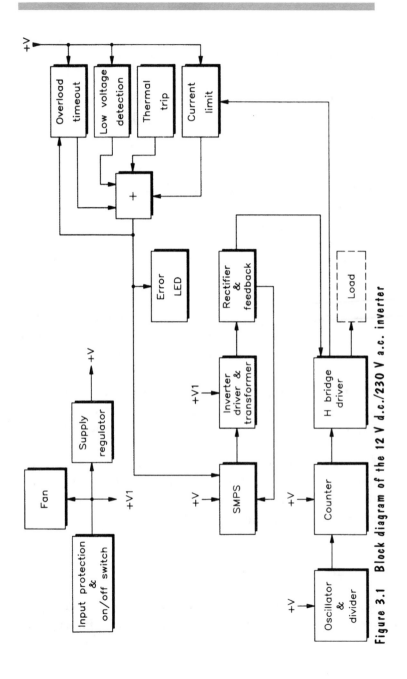

Figure 3.1 Block diagram of the 12 V d.c./230 V a.c. inverter

101

Power supply projects

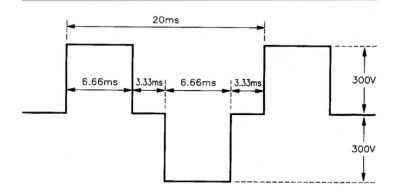

Figure 3.2   Quasi sine-wave

The switch SW1 turns the fan on as well as supplying power to the rest of the circuit. Resistor R45 is required to drop the excess voltage to the fan. Inductor (choke) prevents supply noise generated by the fan from upsetting the regulator VR1, which is required to regulate the supply to the ICs. The zener diode ZD4 is not used on the +12 V d.c. version and is replaced by a wire link.

Capacitors C16 and C6 provide low and high frequency decoupling at the input of the regulator, while capacitors C15 and C7 provide the same function at the output, as well as aiding stability.

The LED LD2 is the power on indicator, resistor R32 is the current limiting resistor.

The heart of the high voltage conversion system is IC4, a dedicated switch mode power supply (SMPS) IC. The frequency of the SMPS is determined by the components

Inverters

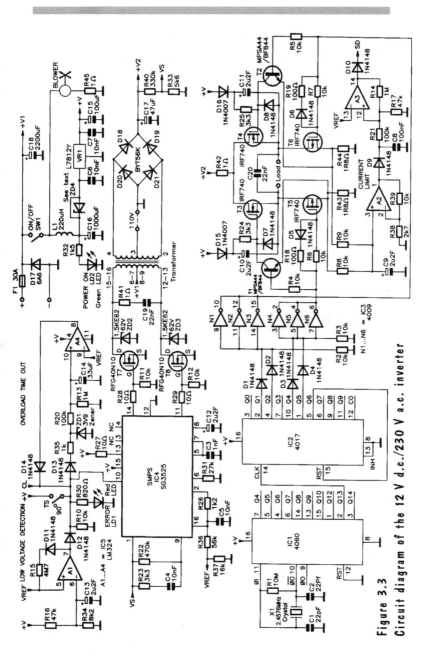

Figure 3.3
Circuit diagram of the 12 V d.c./230 V a.c. inverter

103

Power supply projects

R31 and C3, in this case it is approximately 53 kHz. Pin 16 is the reference voltage pin, and resistors R36 and R37 attenuate the reference voltage to the required level for the rest of the circuit. Pin 10 is the *shutdown input*. Pins 11 and 14 are the outputs which control the MOSFETs T7 and T8 which are the push pull devices for driving the transformer.

Zener diodes ZD2 and ZD3 are there to protect the driver MOSFETs T7 and T8 from the induced e.m.f. produced by the transformer caused by the collapse of the magnetic field. Capacitor C19 and resistor R41 form a Zobel network, which has a multi-function purpose; such as to act as a snubber to slow the rise time at switch on, and to effectively reduce any radiated harmonics, as well as to soften the decay of the magnetic field.

The transformer steps up the high frequency switched input to a high potential. The output of the transformer is then rectified by the bridge rectifier D19 to D21 and smoothed by the reservoir capacitor C17.

The resistors R40 and R33 divide the output potential of the transformer to the required level. This is then fed back to pin 1 of IC4 forming a closed feedback loop; thus allowing the output to be constantly monitored and adjusted as necessary by means of pulse width modulation (PWM), to maintain constant output voltage both on and off load.

The main output, +V2 is connected to the top of the *H bridge driver* via R42; so called because the four output devices T3 to T6) form the vertical lines of the *H* while the load would be the horizontal line.

IC1 is a 4060 14-stage ripple counter/oscillator; the oscillator section also comprises crystal X1, resistor R1 and capacitors C1 and C2. Each section of the ripple counter divides the crystal frequency by two for 13 stages; the final output frequency is 8.192 kHz which is then fed to the clock input of IC2, a 4017 decade counter. The Q6 output of the 4017 is connected to the reset pin to provide the required count. The outputs Q1, Q2 and outputs Q4, Q5 are OR-ed together via the diodes D1, D2 and D3, D4, which feed the parallel 4009 hex inverters N1 to N6 (this is to provide more current).

When the input to the inverters N1, N2 and N3 is low, the FET T6 will be switched on; transistor T2 will also be switched on; which in turn will switch off FET T4. When the input to the inverters N4, N5 and N6 is low, the FET T5 will be switched on; transistor T1 will also be switched on; which in turn will switch off FET T3.

When either of the outputs of Q0 or Q3 are high, the outputs of the inverters N1 to N6 are high; this will turn on FETs T5 and T6, and transistors T1 and T2; and turn off FETs T3 and T4 via T1 and T2; this effectively connects the live and neutral of the load to 0 V via resistors R43 and R44; see Figure 7.2 for the output waveform. The following combination of FETs T3, T4 or T3, T5 or T4, T6 cannot be switched on at the same time.

The current limiting circuit comprises an amplifier to monitor the load current. This operates by using R43 and R44 as current sensing resistors enabling the voltage to be measured and amplified with non-inverting amplifier A2, of which R38 and R39 set the gain. Diode D9 prevents A2 from sinking the voltage on the capacitor C8.

Power supply projects

Important safety considerations:

● never connect the input battery terminals directly to the output,

● do not earth any side of the output to ground,

● do not operate the unit out of its case,

● do not touch the output connections,

● do not extend the battery connection wires,

● do not cover the ventilation holes,

● do not use in wet conditions,

● do observe the same precautions as you would the mains,

● do ensure that power consumption is not more than 250 W,

● do replace fuses with correctly rated ones,

● do place the inverter away from TVs as it may cause interference.

Op-amp A3 is used as a Schmitt trigger, with R14 providing the positive feedback. If the voltage at the non-inverting input is higher than V_{ref}, then the output will latch high, and the output is then fed to the *CL* junction via D10. CL (pin 10 of IC4) is the central connection of all the protection circuits, i.e. the under-voltage detection (A1), overload time out (A4) circuits and the thermal switch (TS). If any protection circuit operates, LD1 will illuminate. In any of these instances IC4 will be disabled by taking pin 10 high.

The low voltage detection circuit also uses a Schmitt trigger based around op-amp A1, with R15 providing the positive feedback for hysteresis.

The inverting input of A1 is connected to the potential divider R16 and R34, capacitor C13 *irons out* any small supply variations. If the voltage at inverting input of A1 drops below V_{ref}, the output will swing high and illuminate LD1.

The overload time out operates in the following manner: should any of the protection circuits trip; capacitor C14 will be charged through R20. The zener diode ZD1 is required to stabilise the charge voltage and therefore the discharge/charge time, determined by R13.

A4 is set up as a comparator trigger with its inverting input connected to V_{ref}. Again, if the non-inverting input is higher than the inverting input; then the output will be high and disable IC4 via pin 10 until capacitor C14 is discharged; which will only happen if the supply is disconnected.

Power supply projects

Construction

It is best to fit the wire links first, and these are marked *J* on the PCB. The two exceptions are 3XJ where three wire links are fitted in parallel and soldered together, and position ZD4 where a link is fitted for the +12 V d.c. version.

Referring to the circuit diagram in Figure 3.3 next fit the resistors, starting with the $^1/_4$ W, $^1/_2$ W, 1 W types. When mounting the two 3 W resistors, stagger them above the board at 10 mm and 15 mm from the PCB to the top of each component.

Next fit the diodes making sure that their orientation is correct. When mounting diodes D18 to D21 there should be 20 mm from the PCB to the top of each component. Do not fit LEDs LD1 and LD2 at this stage.

Locate and fit zener ZD1, making sure that it is correctly orientated on the board. When fitting ZD2 and ZD3, these should be mounted 10 mm from the PCB to the top of each component. A link is fitted in place of ZD4 in the +12 V version.

Identify and fit the inductor: it looks similar to a resistor but has a pale blue or green case.

Next fit the one 14-pin and four 16-pin DIL IC sockets, making sure that the notches on the sockets match those on the PCB.

When fitting the capacitors, observe the correct polarity on the electrolytics. The positive lead is normally longer, and the negative is shown by a series of arrows or – symbols on the casing.

108

Solder the four PCB blade terminals which are marked GND +V and the two *a.c.* onto the PCB. Next identity and mount VR1 the 7812 regulator, making sure that the metal heatsink matches the legend on the PCB. Matching the outlines of T1 and T2 solder these onto the PCB, and then fit the crystal, PCB pins and switch SW1.

Refer to Figure 3.4 for fitting T3 and T4. Note that they are mounted flat on the PCB, and held in position by a 12 mm M3 bolt, shakeproof washer and nut. Figure 3.4 shows also the vertical fitting of T5, T6 and TS (the thermal switch). Note that T5 has its own heatsink and that T6 and TS share a heatsink. Again make sure that the devices match the legend on the PCB.

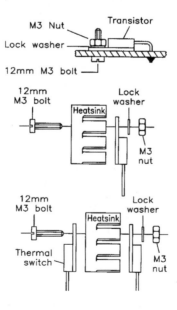

Figure 3.4   T3 to T6 mounting arrangements

Power supply projects

Solder the fuse holder in position and push the fuse F1 into its holder.

Next fit the ICs into the DIL sockets, correctly orientating them so that the notch on the ICs matches those of the sockets.

Assemble the transformer as shown in Figure 3.5. It is formed by fixing together sections of coil formers, cores, and held together by two metal yokes. When the assembly of the transformer is complete, mount it onto the PCB with the markings on it matching those on the board.

Next fit the red and green LEDs LD1 and LD2. These are fitted and soldered in under the PCB, see Figure 3.6 for mounting details.

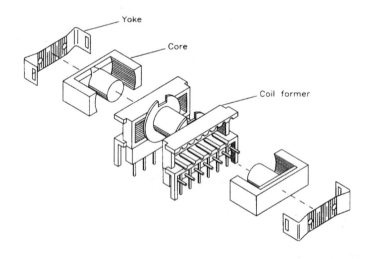

Figure 3.5   Transformer assembly

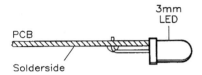

Figure 3.6   LED mounting

Referring to the drawings in Figure 3.7(a) place the two thermally conductive insulators onto the end plate, and using a small amount of thermally conductive paste applied on top, fit T7 and T8. A metal retaining plate can then be bolted to the cover holding T7 and T8 in position by three 15 mm M3 bolts, nuts and shakeproof washers, the order of which is shown in Figure 3.7(b).

Bend the legs of T7 and T8 and solder to the PCB, ensuring that the legs are soldered in the correct place to enable the end plate to fit flat against the case. See Figure 3.7 which shows mounting details.

Assembly

Screw the end plate (with the large hole) to the larger of the two halves of the case. Attach the fan to the end panel using the two M3 30 mm bolts, nuts and shakeproof washers provided. Half slide the PCB into the slot, solder the red (+) and black (–) wires from the fan (blower) to the PCB; the position is marked by (–B+).

Power supply projects

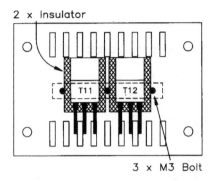

Figure 3.7(a)   Insulators and T11 and T12 mounting on end plate

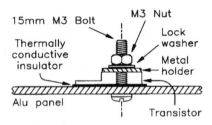

Figure 3.7(b)   Fitting the metal bracket

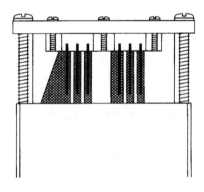

Figure 3.7(c)   Soldering T11 and T12 devices to the PCB

Fit the grommet in place above the power *on* switch.

Solder the blade terminal connectors onto the heavy duty red and black wires. Connect the black wire to *GND* blade terminal and the red to *+V* blade terminal. Feed the free end of the wires through the grommet, and solder the large crocodile clips to the wires. Solder the blade terminal connectors onto the twin mains DS live and neutral wires, and connect to the blade terminals marked *AC*. Feed the free end of the cable through the grommet, and connect a 13 A trailing socket to the mains cable. Now slide the PCB fully home, making sure that the wires passing through the grommet are not too slack and screw the end plate in position, see Figure 3.7(c). Make sure that the leads are not catching and fit the top half of the case and screw in place. Figure 3.8 shows the battery and trailing lead connections. This completes the assembly.

Some devices may offer a capacitive load (computers and some battery chargers), as a result the inverter's heat protection circuit may operate after a short while. It will be necessary to wait a few seconds before resetting the inverter, as it is important to leave the inverter switched on so that the fan can operate.

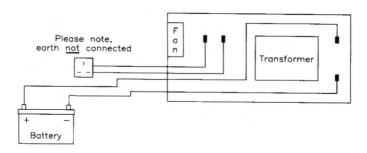

Figure 3.8   Simple wiring diagram

Power supply projects

After the inverter has gone into overload or short circuit protection, it must be reset by switching the unit
off, waiting for 10 seconds and then switching the unit
back on.

During continuous use of the inverter, it is advisable to
leave the vehicle's engine running, to prevent the battery being discharged.

—Take Note —Take Note —Take Note —Take Note —

Important Safety Warning

The 230 V a.c. output from this inverter, and the
high voltages present inside the inverter, are
potentially lethal and must be treated with the
same respect as mains voltage. Every possible
precaution must be taken to prevent the possi-
bility of electric shock. Since this unit does
not provide any connection to the earth pin of
the trailing socket, it must be used only with
Class II double insulated equipment.

Under *no* circumstances should Class I equipment, requiring an earth connection, be
connected to this unit. Under *no* circumstances
should any connection be made to the earth terminal of the trailing socket. Failure to observe
these precautions is likely to expose the user
to electric shock and possible fatal electro-
cution. If in any doubt as to the correct way to
build or use this unit, seek advice from a suitably qualified engineer.

Input voltage:	10.5 to 15 V d.c.
Current consumption:	25 A (max)
Output voltage:	230 V a.c. (± 10%)
Output frequency:	50 Hz (crystal controlled)
Continuous output:	250 W (max)
Peak output power:	500 W
Efficiency:	80% (max)
Dimensions:	270 x 80 x 85 mm
Weight:	1.4 kg

Table 3.1   Specifications

Power supply projects

12 V d.c./230 V a.c. inverter parts list

Resistors — All 0.25 W unless specified

R1	10 M	1
R2–12, 39	10 k	11
R13,14	1 M	2
R15	4M7	1
R16,17	47 k	2
R18,19	100 Ω	2
R20,21	100 k	2
R22	470 k	1
R23–25	3k3	3
R26	1k2	1
R27–29	10 Ω	3
R30	820 Ω	1
R31	27 k	1
R32	1k5	1
R33	5k6	1
R34	8k2	1
R35	1 k	1
R36	56 k	1
R37	16 k	1
R38	2k7	1
R40	330 k (0.5 W)	1
R41,42, 45	1 Ω (1 W)	3
R43,44	1 R 8 Ω (3 W)	2

Capacitors

C1,2	22 pF	2
C3	1 nF	1

C4–7	10 nF	4
C8	100 nF	1
C19,20	22 nF 600 V	2
C9–13	2μ2F 50 V min electrolytic	5
C14	33 μF 16 V electrolytic	1
C15	100 μF 25 V electrolytic	1
C16	1000 μF 35 V electrolytic	1
C17	47 μF 350 V electrolytic	1
C18	2200 μF 16 V electrolytic	1

Semiconductors

T1,2	MPSA44/BF844	2
T3–6	IRF740	4
T7,8	RFG40N10	2
IC1	4060	1
IC2	4017	1
IC3	4009	1
IC4	SG3525A (SMPS)	1
IC5	LM324	1
D1–14	1N4148	14
D15,16	1N4007	2
D17	6A6	1
D18–21	BYT56K	4
ZD1	3V9 zener	1
ZD2,3	62 V 1.5KE62	2
ZD4	wire link (see text)	1
LD1	3 mm LED red	1
LD2	3 mm LED green	1
VR1	L7812 regulator	1

Miscellaneous

	PCB	1
L1	220 μH	1

Power supply projects

TR1	TR3507 transformer yoke	2
	core	2
	coil former	1
TS	90°C thermal switch	1
SW1	SPDT min switch	1
X1	2.4576 MHz crystal	1
F1	30 A fuse	1
	16-pin IC sockets	5
	heatsinks	2
	thermally conductive insulators	2
	+12 V d.c. fan	1
	spade posts	4
	spade connectors	4
	case	1
	M3 12 mm bolts	2
	M3 15 mm bolts	3
	M3 30 mm bolts	2
	M3 nuts	9
	M3 shakeproof washers	9
	metal retainer plate	1

Optional (not in kit)

trailing mains socket	1	(HL73Q)
2-core mains cable	as req	(XR47B)
silicone grease tube	1	(HQ00A)

The above items (excluding Optional) are available as a kit, order as VF35Q

8 W, 12 V fluorescent tube driver

Fluorescent lights have many advantages over incandescent lamps when used out of doors especially when limited power resources are available. Heat output is very low, reducing the risk of fire especially in tents and an average family car battery could supply sufficient power for up to 15 hours continuous use. Light output radiates from the length of the tube, not from one focused point making diffusers and reflectors unnecessary, and being much kinder on the eyes. Unfortunately there is one problem with fluorescent tubes: high voltages are required to *strike* and run them, so a method of driving many hundreds of volts from a 12 volt source must be employed. Our fluorescent tube driver meets the requirements and provides a system at much lower cost than commercially available units.

Circuit description

When power is applied, TR1 is turned on hard via R1 and L2. L1 is energised and passes a high current which induces a pulse in L2 and turns TR1 off for the duration of the pulse. No current flows through L1 at this time and L2 offers a low impedance path from R1 to TR1 base thus turning it on again. Due to this alternating field a large voltage is developed across L1 — around 100 volts — and step-up winding L3 generates several hundred volts, enough to strike the fluorescent tube. The load now remains constant across L3 and the oscillation frequency is maintained by time constant R1 and C2.

Power supply projects

Under normal load running conditions a 50 kHz square-wave at 250 volts should be present across pins 5 and 6. In case of reversed battery connections, D1 prevents damage to both TR1 and battery from occurring, and it will not pass current under these conditions. C1 decouples the supply rails and prevents RF transmission from long battery-lead cables (see circuit diagram, Figure 3.9).

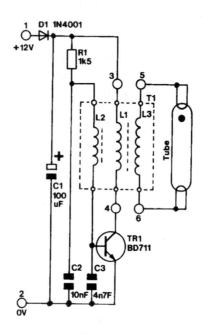

Figure 3.9   Circuit diagram

Transformer construction

Three separate windings are required, see Figure 3.10, these being:

120

- secondary L3: 200 turns of 34 swg (0.3 mm) E/C wire,

- secondary L2: 15 turns of 34 swg (0.3 mm) E/C wire,

- primary L1: 30 turns of 24 swg (0.6 mm) E/C wire.

Wind L3 first on the bobbin (Figure 8.2(a)) by tinning the E/C wire and soldering it to the terminal L3 start. Wrap each turn close to the previous one and build up in layers. Approximately 30 to 32 turns can be made across the former, so six layers should be built up as neatly as possible. Terminate L3 finish as before and insulate the windings with a single layer of PVC insulating tape wrapped tightly around the coil. Next wind L2 (Figure 3.10(b)) starting and terminating on the oppo-

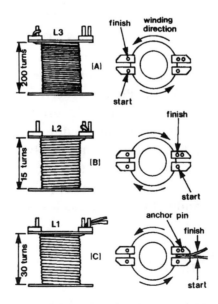

Figure 3.10 Construction of T1

site two bobbin pins (third one not used). Again, spread all 15 turns tightly across the previous coil L3 — eight turns across and 7 turns back. Finally, wind L1 straight on top of L2 (Figure 3.10(c)). Leaving two inches of spare wire, wind two layers, 15 across and 15 back again leaving two inches of spare wire. Wrap three turns of PVC tape tightly around L1 to prevent it from unwinding and drop into one section of T1. Fit the remaining section over the bobbin and secure both halves with metal clips clamped over each end. Before fitting onto the PCB make sure the windings of L2 and L3 have been soldered correctly to their bobbin pins and remove any excess solder which may prevent insertion into the board.

PCB construction

Refer to the parts list and Figure 3.11. Mount the capacitors C2, 3 and resistor R1. Insert diode D1 correctly to the legend on the PCB to ensure correct polarity. Next insert veropins 1 to 6. Position the vaned heatsink and mount TR1 (Figure 3.12) making sure that the leads of TR1 go through the board and tighten the nut and bolt. Insert C1, which is polarised, and finally fit T1. L1 is soldered to pins 3 and 4 and the two wire ends should be scraped to remove the enamel before tinning. Solder components and cut-off all excess leads.

Using the module

Connect an ammeter in series with pin number 1 and +12 volt supply; supply common or −ve goes to pin 2. Set the

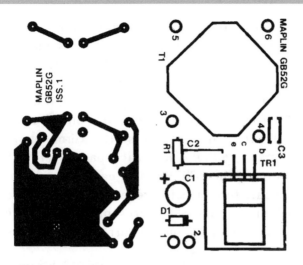

Figure 3.11   PCB legend

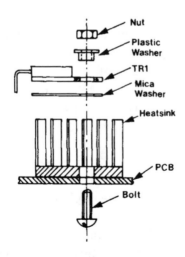

Figure 3.12   Mounting the transformer and heatsink

123

ammeter scale to allow a reading of 1 amp or more and apply power. A high pitch whistling may be heard, with a current reading of 0.4 to 0.5 A. If the reading is 1 A or more, switch off and reverse L1 connections to pins 3 and 4 and check again. Remove power and connect an 8 W 12 inch fluorescent tube across pins 5 and 6. The tube will probably have two starter terminals at each end (four altogether). Join each pair together before connection to the PCB. Keep all connections short and insulate bare terminals to prevent the risk of shock. Remember high voltages are present here and could be dangerous, even with limited current availability!

Apply power again and the tube should glow dimly, then after a second or two light up completely. Check current reading is approximately 0.5 A. No whistling should be audible and the tube should not flicker, but if this is not so, try reversing L1 connections to pins 3 and 4 or reverse tube connections to pins 5 and 6. The inverter can drive two tubes in series (not parallel), at slightly reduced light output levels and the supply current will rise by 100 mA or so when doing this. Resistor R1 can be increased up to 2 k to reduce light output (and supply current) or taken down to 470 R for increased light output, with supply current up to 1 A. With the specified value for R1, tube life expectancy should be high and the prototype has been running for a great many hours without problem.

For housing the tube, clear plastic piping, as used on water tank overflows etc., can be utilised and fitted to a small plastic box containing the inverter. The module could then be potted for safety and a cork fitted into the open end of the pipe.

Tube driver parts list

Resistors — All 0.6 W 1% metal film

R1	1k5 (see text)	1 (M1K5)

Capacitors

C1	100 uF 25 V PC electrolytic	1 (FF11M)
C2	10 nF carbonate	1(WW29G)
C3	4n7F ceramic	1 (WX76H)

Semiconductors

D1	1N4001	1 (QL73Q)
TR1	BD711	1 (WH15R)

Miscellaneous

L1	30 turns x 24 swg ECW	1
L2	15 turns x 34 swg ECW	1
L3	200 turns x 34 swg ECW	1
T1	ferrite pot core type 3	1 (HX09K)
	bobbin type 3	1 (HX10L)
	clip type 3	2 (HX11M)
	24 swg enamelled copper wire	1 (BL28F)
	34 swg enamelled copper wire	1 (BL42V)
	mounting kit	1 (WR23A)
	heatsink	1 (FL58N)

Power supply projects

veropin 2145   1 (FL24B)
tube driver PCB   1 (GB52G)
6BA x $^1/_2$ in bolt   1 (BF06G)
6BA nut   1 (BF18U)
12 V 8 W fluorescent tube   1 (LQ11M)

The above parts are available as a kit, order as
LK35Q

4   Miscellaneous

Intelligent split charge unit

A split charge unit is employed in a car or other similar vehicle to charge an auxiliary (second) battery. An auxiliary battery is often used to power 12 V electrical accessories in a caravan or trailer. The use of such a battery ensures that the towing car's main battery is not discharged. The auxiliary battery can be located in the towing car's boot or in the caravan itself.

Modern caravans are often equipped with a multi-supply refrigerator. Such units can be operated from 230 V a.c. mains, 12 V d.c. or liquefied petroleum gas (LPG). During transit, it is inadvisable for the refrigerator to be operated from LPG for obvious safety reasons; at such times, the refrigerator can be switched to 12 V operation. When the caravan is stationary it can be switched to LPG operation or a.c. mains operation (caravan parks often have electricity hook-up facilities).

Important safety warnings

Before starting installation work, consult the vehicle's manual regarding any special precautions that apply. Take every possible precaution to prevent accidental short circuits occurring as a lead-acid battery is capable of delivering extremely high current. Remove all items of metal jewellery, watches, etc., before starting work. Disconnect the vehicle's battery before connecting the module to the vehicle's electrical system. Please note that some vehicles with electronic engine management systems will require reprogramming by a main dealer after disconnecting the battery. Assuming a negative earth vehicle, disconnect the battery by removing the (–) ground connection first; this will prevent accidental shorting of the (+) terminal to the bodywork or engine. It is essential to use a suitably rated fuse in the supply to this project. For the electrical connections, use suitably rated wire able to carry the required current. If in any doubt as to the correct way to proceed, consult a qualified automotive electrician.

Both *ordinary* car and marine and leisure batteries contain sulphuric acid, produce hydrogen gas and can deliver massive short circuit currents. For these reasons the following points should be observed:

●   follow the manufacturers instructions on installation, charging, use and maintenance,

●   avoid spilling battery acid — place the battery in a purpose-designed enclosure and ensure it is firmly secured. Do not leave a battery unsecured in a car boot; it is liable to topple over and battery acid will spill out. If contact is made with battery acid, flush the affected areas with plenty of cold water. Battery acid will rot clothing and cause severe chemical burns to skin. If battery acid is splashed into eyes, flush with cold water and seek urgent medical attention,

●   hydrogen gas, when mixed with air, is explosive. Avoid sparks, flames or other sources of ignition in the vicinity of the battery. Marine and leisure batteries are often fitted with a vent-pipe to avoid the build-up of hydrogen — follow the manufacturers' instructions on use. Remember hydrogen is less dense than air so it will rise to the highest point in an enclosure.

Power supply projects

The dual split charge unit presented here is able to simultaneously charge an auxiliary battery and supply power to a +12 V d.c. operated refrigerator.

Any old battery?

The choice of auxiliary battery requires some thought and a little knowledge. All too often, people use an ordinary car battery. While a car battery will obviously work, its performance will be disappointing and life expectancy short. What is needed is a marine and leisure battery. Such batteries employ different design and construction to car batteries and cater for quite different requirements.

A car battery is only drained of charge when the car is started, or accessories (lights, radio, etc.) are used without the engine running. When the engine is running, the car's alternator supplies the needs of the car's electrical system and replaces charge lost from the battery during starting. Thus it can be seen, in normal use, the battery is only ever partially discharged and then immediately recharged.

If a car battery is repeatedly discharged until flat then recharged, it will soon lose capacity and become unusable. In a camping and caravanning application, this is the kind of treatment that an auxiliary battery will receive as a matter of course. What is needed is a battery designed for such treatment — a marine and leisure battery is intended to fit the bill.

Marine and leisure batteries are available from several manufacturers in a wide range of sizes/capacities. Smaller batteries will be sufficient for supplying basic lighting needs while a larger one is needed if an electric water pump, colour TV, etc. are to be powered. The exact choice will depend on total demand for power, how long the battery should last before needing to be recharged and how big your wallet is! Good automotive stores, yacht chandlers and camping equipment suppliers will stock a range of marine and leisure batteries, and should be able to offer advice as to the best one to use in a particular application.

Charge!

Many readers will be familiar with nickel cadmium cells/batteries and how they are charged. The method of charging a lead acid battery is quite different. In a car, with the engine running, the alternator provides a constant voltage output of 13.8 V. This voltage is 1.8 V higher than the nominal terminal voltage of a 12 V battery.

By the time a 12 V battery's terminal voltage has fallen to 10 V, it has lost most of its charge. Such a discharged battery will draw current from the alternator. As the battery charges, its terminal voltage will rise to meet the charging voltage. The charging current will correspondingly fall. When the battery reaches full charge, the amount of current drawn will drop to a low level at which point the battery will be continuously trickle-charged and thus maintained at full capacity (lead acid batteries have a slight tendency to self-discharge over a period of time).

Power supply projects

Supply voltage range:	+10 V to +16 V d.c.
Maximum auxiliary battery output current:	10 A
Maximum refrigerator output current:	10 A supply current
Control input:	100 mA maximum
Monitor circuit supply:	50 mA maximum

Table 4.1   Specification

In essence, to charge an auxiliary battery, simply requires it to be connected in parallel with the existing
battery. However, things are not *quite* as simple as that
for a number of reasons:

●   if one battery was discharged and the other was
fully charged, the discharged one would draw current
from the charged one until equilibrium was reached,

●   when current is drawn from the main battery, current would also be drawn from the auxiliary battery. For
this reason, cabling to both batteries would have to be
equally rated for the maximum load (starting),

●   if lights were left on, both batteries would be discharged.

To overcome these problems, a split charging system is
adopted. With such a system, the auxiliary battery is only
connected to the main charging system when the alternator is generating sufficient output to charge the
batteries. This is commonly achieved by using a relay
which is controlled by the charge warning output from
the alternator — normally this output is used to operate
the dashboard charge warning indicator lamp. It is possible however, for problems to be encountered, mainly

132

because the relay can become stuck in the open or closed condition: this will result in the auxiliary battery not charging. With the relay stuck open there is no path for the current to flow. With the relay stuck closed substantial current will flow from the auxiliary battery when the car is started, blowing the split charge circuit fuse, again with the same end result.

The design presented here employs a relay and a high power (30 A) Schottky diode. Together these components ensure that current can only ever flow into the auxiliary battery from the main charging circuit and that charging only takes place when there is sufficient output from the alternator to charge the batteries. The Schottky diode was chosen for its low forward voltage and ability to withstand large current surges.

Circuit description

The split charge unit can be effectively broken down into two main sections: the high current switching circuitry and low current status monitoring circuitry. The overall operation can be seen in simplified form in the block diagram shown in Figure 4.1. The circuit diagram is shown in Figure 4.2 and circuit operation is as follows:

High current circuitry

A high current permanent +12 V feed is supplied from the car's battery to P1 of the split charge unit. Relays RL1 and RL2 switch power to the Auxiliary Battery and Refrigerator Outputs, P2 and P3 respectively — RL3's

Power supply projects

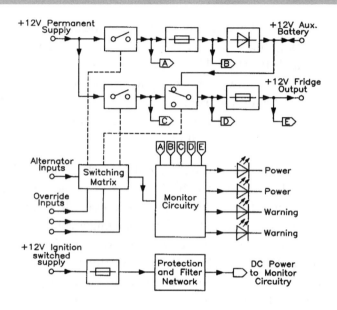

Figure 4.1   Block diagram of the Intelligent Split Charge Unit

presence may be ignored, for the time being at least. RL1 and RL2 are switched by means of a simple diode matrix comprising D4, D5, D7 and D11. These diodes allow either relay to be switched individually or both of them together. D8 and D12 serve to arrest the potentially damaging high voltage spikes produced by the relay coils when they de-energise.

Normally, the Control Input, TB1–2 is connected to the charge warning indicator output from the car's alternator. This output rises from near 0 V, at engine standstill, to (nominally) +13.8 V, when the engine is running and the alternator is providing power to the car's electrical system.

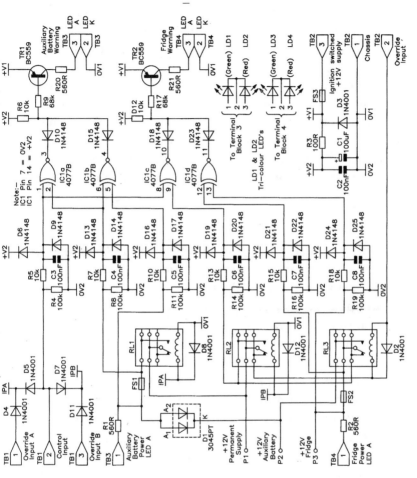

Figure 4.2 Circuit diagram of the Intelligent Split Charge Unit

135

Power supply projects

If the charge warning output is not easily accessible, the Control Input may be supplied from a circuit that becomes live when the ignition circuit is switched on.

When the Control Input is pulled high, diodes D5 and D7 conduct, each of which supply power to the coils of RL1 and RL2 thus causing them to energise. +12 V is then fed to the Auxiliary Battery Output via FS1 and D1 and the Refrigerator Output via FS2. D1, a high current double Schottky rectifier diode ensures that the auxiliary battery cannot discharge by reverse feeding the *car's* electrical system.

It may be desirable to supply power to the Auxiliary Battery or Refrigerator Outputs at times when the engine is not running. This would be desirable when the car is stopped for a short while (at a service station, etc.) to maintain the supply to the refrigerator. Alternatively, the outputs may be used to supply other 12 V accessories (+12 V hand-held spotlight, etc.). Such use is facilitated by Override Inputs A and B, TB1-1 and TB1-3 respectively. Override Input A, when taken to +12 V, switches on RL1 via D4 and supplies power to the Auxiliary Battery Output. Similarly, Override Input B, when taken to +12 V switches on RL2 via D11 and supplies power to the Refrigerator Output. Care should be taken that the car's battery is not drained to such a level that there is insufficient charge to restart the engine. The time for this to happen is a function of the current drawn and the capacity of the battery.

Indication that power is reaching the desired output is given by means of green LEDs connected to TB3-1 (Auxiliary Battery Power Status Output) and TB4-1

(Refrigerator Power Status Output). TB3-2 and TB4-2 provide 0 V/chassis return connections for the LEDs. R1 and R2 serve to limit LED current flow to around 20 mA.

RL3, which has so far been ignored, allows the auxiliary battery supply to be back-fed to the Refrigerator Output when the engine is not running. This may be used for the same reasons as mentioned above, but without discharging the car battery (the current drawn is only that required to operate RL3). Such operation is enabled by taking Override Input C to 0 V/chassis. D2 serves the same function as D8 and D12.

Low current circuitry

Fault detection is provided by means of IC1, a quad exclusive NOR (EXNOR) gate.

IC1 is a CMOS logic device and does not take too kindly to the harsh, electrically noisy, automotive environment. For this reason, each of IC1's inputs is protected from noise and spikes by a low-pass filter and a double diode clamp. Taking pin 1 of IC1(a) as an example, R5 and C3 form the filter and D6 and D9 form the clamp. Each of IC1's remaining inputs are similarly protected. The supply to IC1 is also filtered; R3 and C1 and C2 performing this function. D3 and FS3 protect the monitoring circuit from inadvertent supply reversal. The output from each gate is high when both of its inputs are at the same logic state (i.e. both high or both low) and low when both inputs are in different states (i.e one high and one low). During normal operation the inputs to each gate will be at the same logic level. However, under fault conditions, such as a blown fuse or jammed relay, the respective gate inputs will be at dissimilar logic levels. IC1a monitors operation of RL1 and IC1b monitors the state of F51; their outputs are NANDed by a discrete DTL gate based around TR1 and associated components. Similarly, IC1c monitors operation of RL2 and IC1d monitors the state of FS2; their outputs are NANDed by TR2 and associated components. If either IC1a or IC1b's outputs go low, TR1 will switch on and supply current to the red LED connected to TB3-3 (Auxiliary Battery Warning Status Output). TR2 will switch on in response to a low level output from IC1c or IC1d, supplying current to the red LED connected to TB4-3 (Refrigerator Warning Status Output). TB3-2 and TB4-2 provide 0 V/chassis return connections for the LEDs. R20 and R21 serve to limit LED current flow to around 20 mA.

Construction

Referring to the PCB legend shown in Figure 4.3 and the Parts List, assemble the PCB following normal construction procedures. The Constructors' Guide (XH79L) supplied in the kit offers comprehensive advice on construction techniques. Start by inserting the seven wire links marked *link* on the PCB legend. Next, with the exception of the FS1, FS2 and D1, proceed by fitting the components in ascending size order, i.e. 1N4148 signal diodes, 0.6 W resistors, etc. Care should be exercised with all polarised devices that they are inserted with correct orientation. IC1 is a CMOS device and normal precautions to avoid electrostatic discharge should be observed.

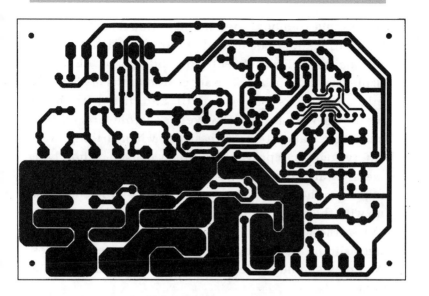

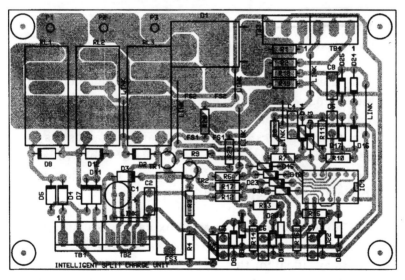

INTELLIGENT SPLIT CHARGE UNIT

Figure 4.3 PCB legend and track

139

Power supply projects

If the suggested box listed in the optional parts list is to be used, it should be drilled as shown in Figure 4.4. Additional holes for cable entry, mounting, etc., should be drilled as appropriate. The suggested box is made of aluminium and is easy to drill and file with basic metal-working tools. However, sensible precautions should be taken to ensure safety when drilling; wear safety goggles to prevent eye injury caused by hot flying metal swarf. Avoid the temptation to remove metal filings from the box by blowing them out — you could end up with them in your eyes. Keep young children and pets out of the work area. Older children who are eager to help should be equipped with suitable safety goggles. All holes should be deburred with a file or a deburring tool.

The suggested box is supplied covered with a protective plastic film, do not remove this until all drilling and filing is complete, otherwise it may become very badly scratched. The drilling details can be marked on the protective film with fine tipped indelible pen. A centre punch should be used on all hole centres, which will help prevent the drill bit skidding. The larger holes and slots can be filed out if suitably sized drill bits are not available, but make frequent checks on progress as it is very easy to file out too much metal!

The final appearance of the box can be substantially improved by spray-painting before assembly. Aerosol car paints are ideal for this and produce excellent results if care is taken. Briefly: rub the box down with extra fine abrasive paper; prime the metal surface with metal primer; apply several thin, even coats of paint and allow to dry thoroughly before attempting assembly. If letter-

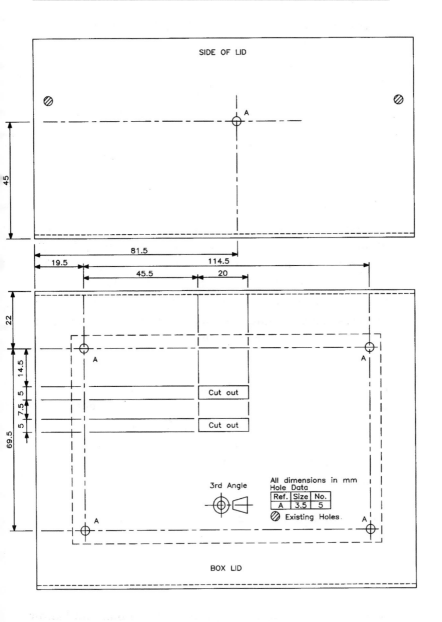

Figure 4.4   Drilling details of the suggested box

ing or graphics are to be used, apply several coats of lacquer to give durability.

FS1 and FS2, which are blade type car fuses, are fitted to the *solder* side of the PCB by means of $1/4$ in blade receptacles. These are fitted as shown in Figure 4.5.

Connect the Schottky rectified diode to the PCB using 10 A wire, as shown in Figure 4.5 — use heatshrink sleeving to cover the exposed connections.

To increase the current rating of the high current PCB tracks it is *vitally important* to apply a *generous* coating of solder to the wide bare tracks not covered with solder resist. Recheck the PCB, ensure that there are no solder splashes bridging any PCB tracks. It is a good idea to spray the PCB with Clear Protective Lacquer (but wait until testing is complete before doing so!), but make sure that the relays and terminal blocks do not suffer lacquer ingress.

Referring once again to Figure 4.5, fit the PCB into the box using M3 x 15 mm threaded spacers. Secure the Schottky diode to the case using M3 hardware and a silicone insulating washer. The device specified has an isolated mounting hole and thus does *not* require an insulating bush — however, the silicone washer *must* be used, otherwise the metal case will short circuit to the diode substrate. Fit the fuses into their holders.

Testing

A bench power supply capable of delivering +12 V d.c. at 200 mA, or a *fuse* protected 12 V supply from a bat-

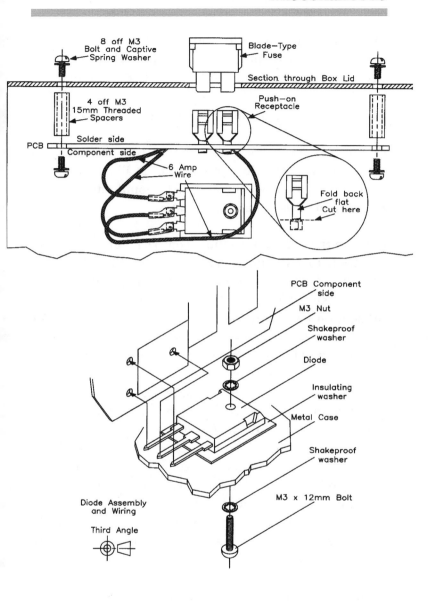

Figure 4.5 Exploded assembly diagram

Power supply projects

tery can be used to test the intelligent Split Charge Unit. Referring to the wiring diagram shown in Figure 4.6, connect four LEDs to TB3 and FB4 — the LEDs connected to TB3-1 and TB4-1 should be green (power) and the LEDs connected to TB3-3 and TB4-3 should be red (warning).

Connect TB2-1 to 0 V and connect P1 to +12 V, at this stage no relays should energise or LEDs illuminate. Connect TB1-2 to +12 V, RL1 and RL2 should energise and both green LEDs should illuminate.

Remove the +12 V connection from TB1-2 and transfer it to TB1-1, RL1 should energise and the green LED connected to TB3-1 should illuminate. Using a multimeter, set to a suitable d.c. volts range, check that +12V is present on P2 with respect to 0 V.

Remove the 112 V connection from TB1-1 and transfer it to TB1-3, RL2 should energise and the green LED connected to TB4-1 should illuminate. Using a multimeter, set to a suitable d.c. volts range, check that +12 V is present on P3 with respect to 0 V.

Remove the +12 V connection from TB1-3 and transfer it to P2, no LEDs should illuminate. Connect TB2-2 to 0 V, RL3 should energise and the green LED connected to TB4-1 should illuminate. Using a multimeter, set to a suitable d.c. volts range, check that +12 V is present on P3 with respect to 0 V.

Without removing any of the previous connections, connect TB1-3 to +12 V, RL2 should energise and RL3 de-energise. The green LED connected to TB4-1 should remain illuminated.

144

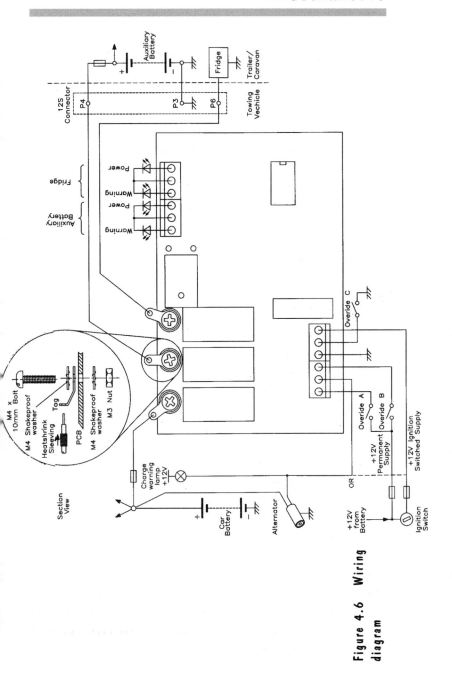

Figure 4.6 Wiring diagram

Power supply projects

Remove the +12 V connections from P2 and TB1-3; remove the 0 V connection from TB2-2. Connect TB2-3 to +12 V, both red LEDs will flash briefly. Connect TB1-2 to +12 V, both green LEDs should illuminate. Remove FS1, the green LED connected to TB3-1 should extinguish and the red LED connected to TB3-3 should illuminate. Refitting FS1 should restore the previous condition.

Remove FS2, the green LED connected to TB4-1 should extinguish and the red LED connected to TB4-3 should illuminate. Refitting FS1 should restore the previous condition.

Carefully remove the covers of RL1 and RL2 by *gently* squeezing the *short* faces of the plastic case and pulling upwards.

Carefully insert a thin piece of paper between the normally open contacts of RL1. The green LED connected to TB3-1 should extinguish and the red LED connected to TB3-3 should illuminate. Removing the paper should restore the previous condition.

Carefully insert a thin piece of paper between the normally open contacts of RL2. The green LED connected to TB4-1 should extinguish and the red LED connected to TB4-3 should illuminate. Removing the paper should restore the previous condition.

Remove the +12 V connection from TB1-3. *Carefully* close the normally open contacts of RL1. The green LED connected to TB3-1 should illuminate and the red LED connected to TB3-3 should illuminate. Releasing the contacts should restore the previous condition.

146

Carefully close the normally open contacts of RL1. The green LED connected to TB3-1 should illuminate and the red LED connected to TB3-3 should illuminate. Releasing the contacts should restore the previous condition.

Carefully refit the plastic cases of RL1 and RL2.

Remove the +12 V connection from P1, both red LEDs should illuminate and both green LEDs should extinguish.

Assuming that the Split Charge Unit performed as described, testing is complete. If results other than those described were experienced, recheck all stages of construction for errors.

Connector/ pin number	Function
TB1-1	Override A input (+12 V to switch)
TB1-2	Control input (+12 V to switch)
TB1-3	Override B input (+12 V to switch)
TB2-1	0 V/chassis
TB2-2	Override C input (0 V to switch)
TB2-3	+12 V ignition switched supply input
TB3-1	Auxiliary battery power status green LED anode
TB3-2	Auxiliary battery status LED cathodes
TB3-3	Auxiliary battery warning status red LED anode
TB4-1	Refrigerator power status green LED anode
TB4-2	Refrigerator status LED cathodes
TB4-3	Refrigerator warning status red LED anode
P1	+12 V permanent supply input
P2	+12 V auxiliary battery output/input
P3	+12 V refrigerator output

Table 4.2 Intelligent Split Charge Unit connector designations

Power supply projects

Installation

Before starting installation, read the warnings given at the beginning of this article. A workshop manual, such as the popular Haynes series, will greatly assist wire tracing. Use a multimeter or a circuit tester to confirm wiring arrangements *before* wires are cut. All connections must be both mechanically and electrically sound. Where wires are joined, solder, crimp or terminal block connectors should be used. All connections should be fully insulated to prevent short circuits.

Find a convenient location for the unit to be fitted, e.g. boot, and secure the box using M4 hardware. Referring to the wiring diagram shown in Figure 4.6 and the 125 connector pin-out diagram shown in Figure 4.7, make the electrical connections.

The status LEDs and override switches should be conveniently located. Care should be exercised when

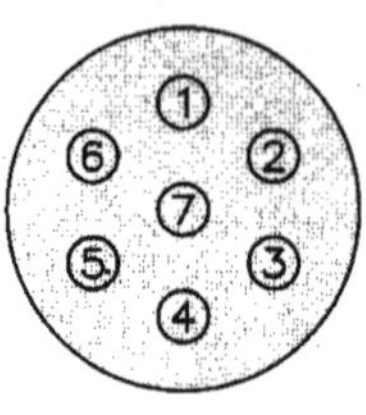

Pin No.	European designation	Function	Wire colour
1	L	Reversing lights	Yellow
2	54G	Spare	Blue
3	31	0 V/chassis	White
4	R	Auxiliary battery	Green
5	58R	Warning light	Brown
6	54	Refrigerator	Red
7	58L	Spare	Black

Figure 4.7 12 V supplementary connector pinout

148

choosing and siting the override switches to prevent accidental operation. The override switches need only be fitted if they are required — they are not essential to the basic operation of the unit.

The high current connection between the car battery and the split charge unit should be made with 30 A wire and include a 20 A fuse fitted as physically close to the battery as possible.

The output connections from the split charge unit to the auxiliary battery and the refrigerator should be made using 10 A wire. The connection to the auxiliary battery should include a 10 A fuse fitted as physically close to the auxiliary battery as possible. The 0 V/chassis connection TB2-1 should be made using 10 A wire.

These protection measures may appear to be excessive, but are required to avoid excessive current flow if accidental short circuits occur.

The remainder of the connection can be made using low current hook-up wire, multicore *burglar alarm cable* is ideal. The cathodes of the status LEDs may alternatively be returned to car chassis instead of TB3-2 and TB4-2 if this saves wiring. The control input TB1-2 may be connected to either the charge warning output from the alternator *or* the ignition switched +12 V supply (*not both*); the alternator's charge warning output is preferable but it may not always be accessible or suitable. In case of doubt or problems, use the ignition switched +12 V supply. *Do not* use the main charge output from the alternator, otherwise the split charge unit will be permanently enabled. The +12 V power input TB2-3 *must only* be connected to the ignition switched +12 V supply and *not* the alternator charge warning output. Override Inputs A, TB1-1, and B, TB1-3, should be connected via independent switches to the +12 V permanent supply. Override input C should be connected via an independent switch to the car chassis.

Double-check *all* connections before reconnecting the car battery.

Use

With the ignition switched off and all override switches off (if fitted) all the status LEDs should be extinguished.

Alternator controlled

Turn on the ignition, both red warning status LEDs will briefly flash. Start the car, when the charge warning light extinguishes, both green power status LEDs will illuminate. Turn off the ignition and both green power status LEDs will extinguish. Similarly, if at low engine revs the charge warning light illuminates (or the engine stalls), both green power status LEDs will extinguish.

Ignition controlled

Turn on the ignition, both green power status LEDs will illuminate. Turn off the ignition and both green power status LEDs will extinguish.

Warning indication

If the supply fuse to the Split Charge unit blows, or the fuses within the Split Charge Unit blow, the appropriate red warning status LEDs will illuminate when the engine is running (alternator controlled) or the ignition is switched on (ignition controlled). Similarly, if a relay becomes stuck open or closed, the appropriate red warning status LED will illuminate. If a relay becomes stuck open, the appropriate green power status LED will not illuminate. If a relay becomes stuck closed, the appropriate green power status LED will permanently illuminate.

Override switches

Override switch A allows power to be fed to the auxiliary battery output when the engine is not running/ignition is switched off.

Override switch B allows power to be fed to the refrigerator output when the engine is not running/ignition is switched off.

Note: care should be exercised when using override switches A and B that the car battery is not drained to the point where the car will not start.

Override switch C allows power to be back-fed from the auxiliary battery to the refrigerator output when the engine is not running/ignition is switched off.

Power supply projects

To check that the auxiliary battery is connected and its fuse is intact, operate override switch C with the ignition switch off; the *refrigerator* power status LED will illuminate if all is well.

Intelligent split charge unit parts list

Resistors — All 0.6 W 1% metal film (unless specified)

R1,2, 20,21	560 Ω	4	(M560R)
R3	100 Ω	1	(M100R)
R4,8,11, 14,16,19	100 k	6	(M100K)
R5,6,7, 10,12,13, 15,18	10 k	8	(M10K)
R9,17	68 k	2	(M68K)

Capacitors

C1	100 µF 25 V radial electrolyic	1	(FF11M)
C2-8	100 nF 50 V disc ceramic	7	(BX03D)

Semiconductors

D1	MBR3045PT	1	(GX38R)
D2-5,7,8, 11,12	1N4001	8	(QL73Q)
D6,9,10, 13-25	1N4148	16	(QL80B)
TR1,2	BC559	2	(QQ18U)
IC1	HCF4077BEY	1	(QW47B)

Power supply projects

Miscellaneous

RL1–3	12 V 16 A relay	3	(YX99H)
	20 mm fuseholder	1	(DA61R)
	T100 mA 20 mm fuse	1	(WR00A)
	push-on receptacle	1	(HF10L)
	10 A blade fuse	2	(KU21X)
	3-way PCB terminal block	4	(JY94C)
	14-pin DIL socket	1	(BL18U)
	M3 x 12 mm steel bolt	1	(JY23A)
	M3 shakeproof washer	1	(BF44X)
	M3 steel nut	1	(JD61R)
	TO3P silicone insulator washer	1	(UK86T)
	10 A wire red	1	(XR36P)
	3.2 mm diameter heatshrink sleeving	1	(BF88V)
	4.3 mm insulated tag	1	(JH71N)
	M4 x 10 mm steel bolt	1	(JY14Q)
	M4 steel nut	1	(JD60Q)
	M4 shakeproof washer	1	(BF43W)
	PCB	1	(GH82D)
	instruction leaflet	1	(XU77J)
	constructors' guide	1	(XH79L)

Optional (not in kit)

	M3 x 15 mm threaded insulated spacer	1	(FS37R)
	30 A wire red	as req	(XR59P)
	30 A wire black	as req	(XR57M)
	10 A wire red	as req	(XR36P)
	10 A wire black	as req	(XR32K)

4-way low current cable	as req	(XR89W)
6-way low current cable	as req	(XS54J)
8-way low current cable	as req	(CW70M)
4.3 mm insulated tag	1	(JH71N)
aluminium box	1	(XB69A)
grommet	1	(JX65V)
M4 x 10 mm bolt	1	(JY14Q)
M4 nut	1	(JD60Q)
M4 shakeproof washer	1	(BF43W)
5 mm red LED	2	(WL27E)
5 mm green LED	2	(WL28F)
5 mm LED clip	4	(UK14Q)
20 A blade fuse	1	(KU23A)
10 A blade fuse	1	(KU21X)
T100 mA 20 mm fuse	1	(WR00A)
in-line fuseholder	1	(RC71N)

The above items (excluding Optional) are available as a kit, order as LT60Q

Power supply for a valve amplifier

This project found its birth as a PSU for the Maplin Newton all-valve pre-amplifier. While valve amplifiers aren't common these days, there's always someone wanting one. While specifically for the Newton, the PSU here *could* be used by the enthusiastic reader for other valve-based applications.

A word of warning, however, before you start.

● *you must* only use parts from the kit, and build it strictly according to the instructions,

● *you must* take particular care with wiring up the HT power supply, chassis earth and common signal earth connections. Follow the accompanying wiring diagrams implicitly,

● *you must not* attempt to make your own PCBs — you can try *hardwiring* in the good old fashioned way with tag boards and the like, *if you have the relevant experience*, otherwise always use the ready-made PCB that includes a solder resist layer as an aid to insulation and user safety.

The power supply unit

The PSU module produces three different outputs. These are the main HT supply common to all valve circuits, a conventional 6.3 V a.c. heater supply, and a special 12.6 V regulated d.c. heater supply specifically for valves which require it. (See also Tables 4.3, 4 and 5).

Miscellaneous

Transformer core material:	Low-field grain-oriented steel
Primary windings:	2 x 115 V (dual UK/US standard)
Secondary windings:	3
HT voltage:	350 V max*
HT current:	100 mA max*
Heater #1 voltage:	6.3 V
Heater #1 current:	1.5 A max
Heater #2 voltage:	15 V
Heater #2 current:	1.5 A max
Mechanical fittings:	4-bolt top cover and frame
Fixing centres:	64.5 x 52 mm (4 off)
Overall dimensions:	80 x 66 x 70 mm high

*Not available simultaneously.

Choke core material:	Electrical steel
Lamination distribution:	Insulated E and I groups
Specific inductance:	7 to 10H nominal at 50 mA d.c.
Maximum d.c. (wire rating):	100 mA
D.C. resistance:	150 Ω approx
D.C. voltage drop:	7.5 V at 50 mA
Mechanical fitting:	2-hole *clamp* type
Fixing centres:	67 to 75 mm (M4 or 4BA)
Overall dimensions excluding lugs:	60 x 52 x 49 mm high

Table 4.3   Specification of PSU transformer and choke

Input voltage:	230 to 240 V a.c. at 50 Hz, or 115 to 120 V at 60 Hz
Primary side protection:	500 mA (UK) or 1 A (US) *quick-blow* Fuse, inline filter and noise suppression
HT output voltage:	<300 to 350 V max (dependent on load)
HT output current:	Up to 50 mA nominal, 100 mA max
HT ripple:	<100 mV peak at 50 mA
HT reservoir discharge method:	Leakage resistor
Reservoir discharge time:	1 minute approx
HT protection:	100 mA *quick-blow* fuse
A.C. heater supply:	6.3 V at 1.5 A max
D.C. heater supply:	12.6 V at 500 mA max

Table 4.4   Specification of Power Supply Module

Power supply projects

Pin number	Function
P1	250 V a.c. in #1
P2	250 V a.c. in #2
P3	6.3 V a.c. in #1
P4	6.3 V a.c. in #2
P5	15 V a.c. in #1
P6	15 V a.c. in #2
P7	HT stage 1 out to L1
P8	L1 in to HT stage 2
P9	+350 V d.c. HT output
P10	HT 0 V and a.c. heater common OVE
P11	6.3 V a.c. heater #1
P12	6.3 V a.c. heater #2
P13	+12.6 V d.c. heater output
P14	d.c. heater 0 V (−)
P15	chassis earth

Table 4.5   PSU module PCB pin designations

Figure 4.8 shows the complete PSU circuit diagram. Many of the components are contained on the single PCB, except a couple that are hardwired to T1 (TS1 and C11), and the choke L1.

Mains transformer T1 has a *dual-standard* primary voltage capability, depending on how its two primary windings are configured. Hence the PSU can be built for the UK/European mains standard of 220 to 240 V a.c. (nominally 230 V a.c.) 50 Hz, or for the USA/North American mains standard of 110 to 120 V a.c. (nominally 115 V a.c.) 60 Hz. Mains power enters the PSU chassis via SK1, a fused and filtered Euro-style mains inlet socket, and is switched by double-pole neon switch S1. Optional Euro-facility mains outlets may be added, connected on one

158

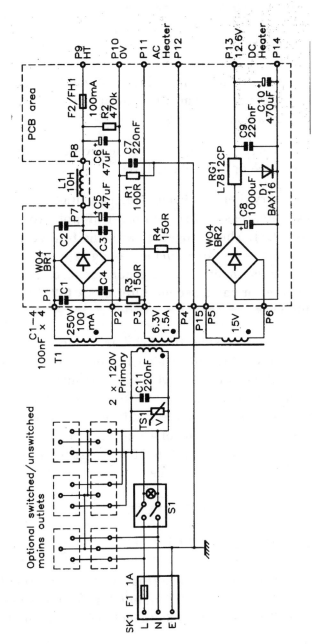

Figure 4.8 PSU circuit diagram

Power supply projects

side or other of S1 depending on whether unswitched or switched outlets to ancillary equipment are required. With the choke in place there may be space on the rear panel of the chassis for up to four such sockets, however, do n*ot* attempt to install these if you have had no experience of fitting them or you are at all unsure of your proficiency at mains wiring!

If you do use them then also use the protective insulating *boots* (see optional Parts List).

At T1 primary there is also included a voltage transient or *spike* suppresser, TS1, and a 220 nF polypropylene capacitor C11. These, together with the filtered mains inlet, may seem like *overkill* but having experienced excessively noisy mains they are not out of place. (The prototype even has a second filter block following the inlet filter).

On the secondary side, the 250 V a.c. winding feeds a bridge rectifier block BR1, which includes noise suppression capacitors C1 to C4, developing unregulated d.c. HT on the main reservoir C5. All these components are on the PCB, but a break is made at this point to include the choke, L1, the output side of which goes to C6; together these two form an integrator or LC low-pass filter to minimise the 100 Hz ripple voltage.

Return of the LF choke

For non-critical audio usage (that is, it is not a test instrument, for example) *HT stabilisers* are not strictly necessary, and if there is sufficient supply decoupling

160

then any level fluctuations manifest themselves as very slowly changing transitions at the output, at a subsonic level below the AF range. If signal coupling capacitors are just small enough in value then not even much of this will be seen at the output.

Hence, it is only really necessary to reduce the ripple, which can be quite a nuisance, but at the same time not lose any more HT level than we can help, it being rather hard to come by as it is, given the unavoidable losses in the transformer.

In the valve heyday, a typical supply electrolytic was the *double-plate* type, effectively two capacitors in one can sharing a common negative side, and between the anodes a choke was connected. The arrangement was very common and employed LF chokes in large numbers. In Figure 4.8, the HT supply from C5 is carried to C6 via the choke L1; the choke is chassis mounted and connected in circuit by wire to PCB pins P7 and P8. If the 100 Hz ripple (a rounded, ramp shaped waveform) on C5 is 10 to 12 V peak (typical), then it is reduced to the order of approximately 70 mV (at least <100 mV) after passing through L1 to C6. This is a reduction factor of 100 times (40 dB) minimum, yet, with a 50 mA drain from C6, the total voltage drop across L1 is only 7.5 V d.c.

Now much less noisy, the HT at C6 is available from pin P9 via protective fuse F2, housed in a covered, PCB mounting fuseholder FH1. R2 is a safety resistor ensuring that the HT line is discharged in the absence of a load or if F2 goes open circuit. The HT supply common earth is at P10, and this point is connected to chassis or mains earth at P15 via R1 and C7. R1 is a hum-*loop block* if other audio equipment connected to the *Newton* shares a mains earth with a signal earth via the mains lead and the screens of audio leads, the loop being completed at the PSU module. R1 is of sufficiently low value to blow F2 should a short circuit occur in the HT supply connections.

The 6.3 V a.c. heater supply is available at pins P11 and P12. A winding centre-tap is emulated by resistors R3 and R4, tying it to 0 V (P10). The heater supply thus forms two opposing, equal and opposite waveforms of 3.15 V each balanced either side of 0 V. When the heater supply wires are formed into twisted pairs, the opposing electric fields cancel, and these techniques reduce hum injection into sensitive areas to a minimum.

Even this may not be quite enough for the very sensitive phono pre-amplifier stages, so a better approach is adopted, exploiting a modern regulator IC. The valve heaters can be operated in series, so a smoothed and regulated 12.6 V d.c. supply is provided by BR2, C8, RG1, C9 and C10, from a 15 V a.c. winding of T1. Designed for 12 V, the output of RG1 is raised to 12.6 V by inserting diode D1 in its common (0 V) connection. This supply is self-contained and electrically isolated from the others, meant to behave, as far as is possible, like a noise-free battery connected to the phono module only. More details about this will emerge later on.

Warning! Before proceeding with any kind of work on this circuit, take heed — high voltages *can kill! Never* touch any high-voltage part of the circuit with either fingers or uninsulated tools unless the power is *off!* While power is on, you should only touch any part of a circuit with an insulated test probe when required. Every time you switch off, adopt the following industrial safety procedure, known by the acronym *side*, which spells out the following steps:

● *switch off* — switch off the main PSU front panel rocker switch, and switch off at the mains outlet wall socket,

● *isolate* — pull the mains lead out of the mains inlet socket at the back of the PSU,

● *discharge* — discharge the main line HT reservoir capacitor to zero volts (*not* with a screwdriver!),

● *earth* — earth the main line HT to chassis 0 V with a leakage resistor to prevent any electrolytics recovering a charge from their own dielectric absorption.

In the design of the PSU *discharging* and *earthing* is automatically taken care of by R2 in the PSU circuit. Please note that it may take the resistor up to one minute to completely discharge the unloaded HT to 0 V. To make doubly sure, you *must* test the main line HT with a multimeter set to high d.c. volts before touching any part of any circuit. This shall hereon be referred to as *the side procedure. Don't cut corners!*

PSU construction

The PCB should be assembled with reference to Figure 4.9, the PCB legend and the Parts List.

Power supply projects

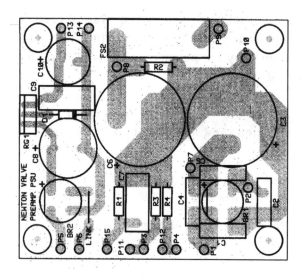

Figure 4.9 Power supply unit PCB legend

Regulator IC RG1 is a type having a plastic mounting tab instead of a metal one. This will make assembly into the chassis much easier as there is no need for a mica insulating kit. While the metal tab would be at earth potential if it were used, and not need an insulating kit anyway, for the reasons discussed earlier the d.c. heater supply must remain electrically isolated from the chassis. This is mentioned in case you were thinking of substituting RG1 with a different device.

Double-check the PCB for the quality of solder joints and correct orientation of components. Once the PCB is installed in the chassis it will be quite awkward to remove again to correct errors! The PCB includes a solder resist on the track side. After removing flux with a PCB cleaner, track side solder joints should be covered with a

164

conformal coating to help the solder resist prevent creepage, or tracking, between points of high potential difference. Some areas are at the full HT line potential.

Preparing the chassis

Cutting and drilling details are given in Figure 4.10. All holes are made in the main body of the 8 x 6 inch aluminium chassis; the removable lid will become the bottom, not the top!

Assembling the chassis

With all holes prepared, begin by mounting T1 with reference to Figure 4.11. To comply with Class 1 requirements for mains powered equipment, we must ensure that the top cover of T1 is satisfactorily earthed to the chassis metalwork on fitting. It is not sufficient to rely on a metal-to-metal contact. One bolt should have its fibre washer replaced with an M5 shakeproof washer. Also place the rectangular steel frame over the lower side of the former, carefully manoeuvring it over the solder tags, until it seats onto the core.

Supporting the chassis on its side with T1 in situ (don't let it move!), place three of the fibre washers, plain washers, shakeproof washers and nuts (supplied in a plastic bag with the transformer) only onto those bolts also having fibre washers at the top, and tighten lightly. For the bolt having the shakeproof washer at the top, replace its fibre and plain washers, then add the M5 solder-tag

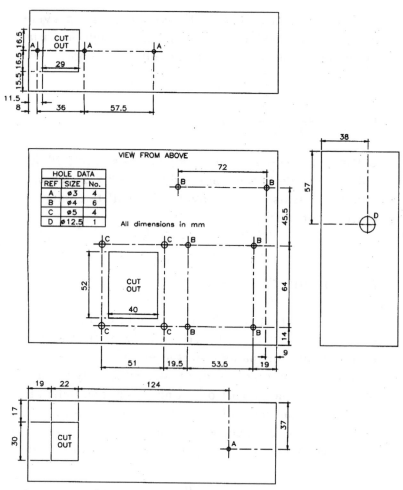

Figure 4.10   PSU chassis drilling details

washer beneath the shakeproof washer and nut. This
should be one of the two bolts near the centre of the
chassis. Eventually an earth strap will connect this to
the chassis; by having shakeproof washers at the top and
bottom, this bolt will ensure that the top cover is electrically earthed.

166

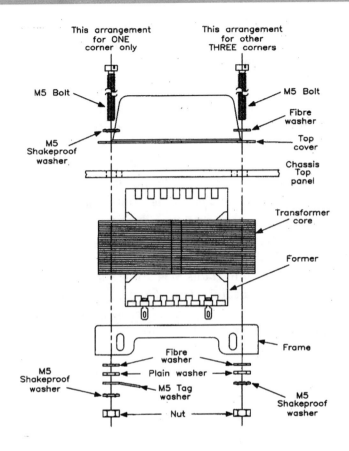

Figure 4.11   Mounting the mains transformer

This done, install the mains rocker switch by pressing it into its rectangular cut-out (all terminals orientated towards what will be the top of the chassis) then the fused Euro mains inlet socket at the rear (fuse tray towards what will be the bottom of the chassis). Secure in place with two M3 x 10 mm bolts, shakeproof washers and nuts.

Power supply projects

Mount the four M4 x 14 mm threaded spacers to the inside of the chassis if not already in place, using the two M4 x 10 mm screws through the top panel.

Mount the choke to the rear corner of the top panel using M4 x 10 mm bolts, shakeproof washers and nuts as in Figure 4.12. The finished PCB can be installed onto its four mounting pillars. In so doing, carefully bend out the leads of RG1 so that it is flat against the inside of the front panel. Its fixing hole should be lined up with the M3 clearance hole drilled in the front panel; if this is not possible file out or re-drill the chassis hole until they do line up. If a separate aluminium front panel is fitted, this should be drilled at this position accurately. RG1 is retained with a countersunk M3 screw, shakeproof washer and nut; the separately fitted front panel should have a countersunk hole such that the screw head is flush with the surface. Thereafter it can be hidden with filler and paint or a stick-on front panel label. *Do not* over-tighten! The PCB is secured with four M4 x 6 mm bolts to the pillars.

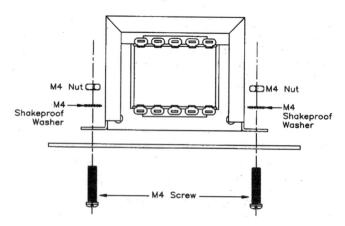

Figure 4.12   Mounting the choke on the chassis

Mains wiring

Complete the mains side wiring with reference to Figure 4.13(a). Prepare a 23 cm length of green/yellow power connection wire with a $^1/_4$ in push-on connector crimped on one stripped end (no insulating sleeve required), and push the connector onto the $^1/_4$ in earth terminal of SK1.

Prepare one blue and one brown 15 cm length of power wire with insulated $^1/_4$ in push-on connectors at each end, and include the insulating boot for SK1. Prepare a further blue and brown pair, 12 cm long, with insulated connectors at one end of each only.

Referring to Figure 4.13(a), connect the Live and Neutral terminals of the Euro mains inlet socket SK1 with the lower (as you see them from the bottom of the chassis) Live and Neutral terminals of the rocker switch SW1 (not the central pair), using the brown and blue 15 cm leads. Also push the green/yellow earth lead from the earth terminal of SK1 through the boot, and cover all connections of SK1 with the boot. The boot should be stretched over the rear end of the metal body of SK1, and may need persuading with a thin-bladed screwdriver or similar, and perhaps a little lubricant. (It may be a good idea to press the boot onto SK1 as soon as you receive the kit to *train* it into the right shape before final assembly.). Likewise connect the brown and blue 12 cm leads to the central terminals of SW1.

Transformer primary connections

The primaries of T1 are wired according to the country of use, that is for UK/European or USA/North American operation:

Power supply projects

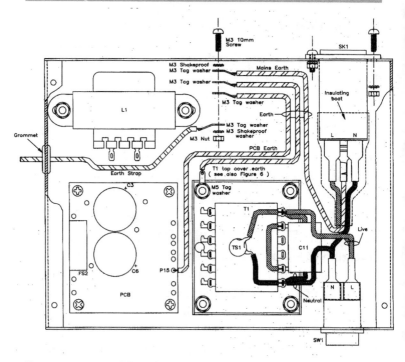

Figure 4.13(a) PSU mains side wiring including configuraton of T1 primaries for UK/European mains standard

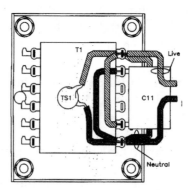

Figure 4.13(b) Configuring for US/North American mains standard

170

● for UK/European use, join the centre two primary tags on T1 with a single short length of power wire, as shown in Figure 4.13(a),

● for USA/North American use, connect the primaries in parallel using two short lengths of power wire as in Figure 4.13(b).

Simultaneously hardwire the yellow 220 nF polypropylene capacitor C11, with the transient suppresser TS1, across the outer pair of primary tags, locating them as shown in Figure 4.13(a) (or Figure 4.13(b)). Both should have their leads sleeved with insulation stripped from power connection wire,

● for use in the UK/Europe, TS1 should be the 250 V type,

● for use in the USA/North America, TS1 should be the 130 V type.

Strip and solder the loose ends of the 12 cm leads from SW1 to the same outer mains primary solder tags of T1; trim lengths to suit.

Earth wiring

Solder an M3 solder-tag washer to the loose end of the green/yellow wire from SK1. Prepare 18 and 6 cm lengths of green/yellow power wire with M3 solder-tag washers at one end only. Prepare a 20 cm length of green/yellow power wire with M3 solder-tag washers at both ends.

Power supply projects

Connect all four wires to the common earthing point on the rear panel using an M3 x 10 mm bolt and nut. Two shakeproof washers should be situated on either side of the stack of four solder tags, that is at the chassis surface and beneath the nut. The *free* earthing strap is to connect the PSU and the pre-amplifier chassis together electrically via the interconnecting grommet when they are joined, and is attached to each via dedicated fixings, using M3 x 10 mm bolts, shakeproof washers and nuts at each tag washer.

Finally install the fuse F1 into the fuseholder tray of SK1 (the tray is released by squeezing the clips at either side), the value of which is according to the country of use:

- for UK/European use, fit the F500 mA ceramic fuse,

- for USA/North American use, fit the F1 A ceramic fuse.

Secondary side wiring

Referring to Figure 4.14, connect six lengths of brown, stranded hook-up wire to the secondary tags of T1 (approximately 4 cm long). Cover each tag with a $^1/_2$ in length of red, heat-resistant sleeving. Connect to the PCB as follows: the 15 V pair to P5 and P6; 6.3 V pair to P3 and P4; 250 V pair to P1 and P2.

Connect the choke tags to P7 and P8 with brown hook-up wire, with heat-resistant sleeving over the choke solder tags. With this the PSU is complete.

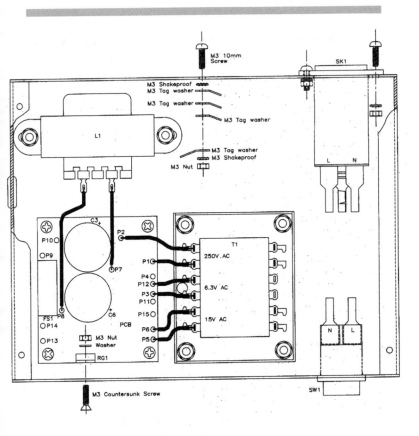

Figure 4.14   PSU secondary side wiring

Testing the PSU

With the chassis still upside down and the bottom cover
off, plug a Euro mains lead into SK1, switch on at the
mains socket and switch on the front panel rocker SW1.
The red neon lamp should light and the transformer may
be heard to hum slightly.

Power supply projects

Set a multimeter to its highest (i.e., 500 to 1000 V) a.c. voltage range, and with insulated probes check for 250 V a.c. (approximately) at T1 secondary output pins P1, P2 on the PCB. Remove probes. Switch to 10 V a.c. range or equivalent, and check heater supply output across pins P11 and P12, it should be 6.3 V between both pins, and 3.15 V between either and the common earth 0VE P10 on the PCB.

Switch to a high d.c. volts range (500 to 1000 V), and test the main line HT output against 0VE (black probe to P10, red probe to P9). It should be approximately 350 V d.c.

(In use the HT level is reduced, due to internal winding resistance in the T1 HT secondary, to approximately 300 V.) Remove probes.

Switch off at the front panel. Stand by with the multimeter probes and recheck the HT level. It should be falling; this proves that the safety discharge resistor R2 is working. If you need to sort out a problem, carry out the complete *side* procedure *before touching anything!* It will take nearly a minute for the HT to completely discharge, in the absence of any other load.

If all is well after the above tests, switch off at the mains socket and remove the mains lead. Apply both the Mains Warning and High Voltage Warning labels to the bottom cover, and temporarily fit it to the chassis with four of the self-tapping screws.

Newton valve pre-amp PSU parts list

Resistors — All 0.6 W 1% metal film

R1	100 Ω	1 (M100R)
R2	470 k	1 (M470K)
R3,4	150 Ω	2 (M150R)

Capacitors

C1,2,3,4	10 nF 500 V high voltage disc ceramic	4 (BX15R)
C5,6	47 µF 450 V radial electrolytic	2 (JL18U)
C7,9	220 nF mylar film	2(WW83E)
C8	1000 µF 35 V radial electrolytic	1 (FF18U)
C10	470 µF 35 V radial electrolytic	1 (FF16S)
C11	220 nF 1000 V polypropylene (class X/Y)	1 (FA22Y)

Semiconductors

D1	BAX16	1 (QB29G)
RG1	L7812CP	1 (CR16S)
BR1,2	WO4	2 (QL40T)

Miscellaneous

TS1	250 V a.c. transient suppressor	1 (HW13P)
TS1	130 V a.c. transient suppressor	1 (CP75S)
L1	10H 100 mA choke	1 (ST28F)
T1	115 V/230 V to 350 V/15 V/6.3 V transformer	1 (ST29G)

175

Power supply projects

FS1	F500 mA 20 mm ceramic fuse	1	(DA05F)
FS1	F1 A 20 mm ceramic fuse	1	(DA06G)
FS2	F100 mA 20 mm glass fuse	1	(WR00A)
	PCB fuseholder and cover	1	(KU29G)
S1	dual red neon rocker switch	1	(YR70M)
SK1	fused inlet/filter	1	(KR99H)
	cover for fused inlet/filter	1	(JK67X)
	AC86 aluminium chassis	1	(XB68Y)
	$^1/_4$ in push-on receptacle	1	(HF10L)
	$^1/_4$ in push-on receptacle covers	1	(FE65V)
	9.5 mm grommet	1	(JX63T)
	mains warning label	1	(WH48C)
	HV warning label	1	(DM55K)
	6 A green/yellow wire	1	(XR38R)
	6 A brown wire	1	(XR34M)
	6 A blue wire	1	(XR33L)
	1.4 A (10 m) brown wire	1	(BL02C)
	red heat-resistant sleeving	1	(BL70M)
	1 mm PCB pin single ended	1	(FL24B)
	M3 x 10 mm steel bolt	1	(JY22Y)
	M3 x 10 mm countersunk bolt	1	(LR57M)
	M3 steel nut	1	(JD61R)
	M3 shakeproof washer	1	(BF44X)
	M4 steel nut	1	(JD60Q)
	M4 x 10 mm steel bolt	1	(JY14Q)
	M4 x 6 mm steel bolt	1	(JY13P)
	M4 x 14 mm threaded spacer	1	(FG39N)
	M5 solder tag	1	(LR62S)
	M3 solder tag	1	(LR64U)
	M4 shakeproof washer	1	(BF43W)
	PCB	1	(GH98G)
	instruction leaflet	1	(XV10L)
	constructors' guide	1	(XH79L)

Optional (not in kit)

Euro outlet	as req	(HL42V)
insulating cover for		
euro outlet	as req	(JK69A)

The above parts (excluding Optional) are available as a kit, order as LT75S